The Mystery of Life's Origin

The Mystery of Life's Origin:

Reassessing Current Theories

Charles B. Thaxton
Walter L. Bradley
Roger L. Olsen

Philosophical Library
New York

The authors have prepared this book in conjunction with the publishing program of the Foundation for Thought and Ethics.

Library of Congress Cataloging in Publication Data

Thaxton, Charles B.
The mystery of life's origin.

Bibliography: p. 220
Includes index.
1. Life—Origin. 2. Chemical evolution. 3. Life—Origin—Research. 4. Chemical evolution—Research.
I. Bradley, Walter L. II. Olsen, Roger L. III. Title.
QH325.B68 1984 577 83-17463
ISBN 0-8022-2446-6 (Hardbound)
ISBN 0-8022-2447-4 (paperback)

Published 1984 by Philosophical Library, Inc.,
200 West 57th Street, New York, New York 10019.

Manufactured in the United States of America.

Foreword

The Mystery of Life's Origin presents an extraordinary new analysis of an age-old question: How did life start on earth? The authors deal forthrightly and brilliantly with the major problems confronting scientists today in their search for life's origins. They understand the impasse in current laboratory and theoretical research and suggest a way around it. Their arguments are cogent, original, and compelling. This book is sure to stimulate much animated discussion among scientists and laymen. It is very likely that research on life's origins will move in somewhat different directions once the professionals have read this important work.

The modern experimental study of the origin of the first life on earth is now entering its fourth decade, if we date the inception of this field of research to Stanley Miller's pioneering work in the early 1950s. Since Miller's identification of several (racemic) protein-forming amino acids in his electric discharge apparatus, numerous follow-up studies have been conducted. Conforming in varying degrees to the requirements of the so-called "simulation paradigm," these experiments have yielded detectable amounts of most of the major kinds of biochemical substance as well as a variety of organic microscopic structures suggested to be similar to the historical precursors of the first living cells.

This program of research can be regarded as a natural extension of Darwin's evolutionary views of the last century. The goal of the work is to find plausible uniformitarian mechanisms for the gradual spontaneous generation of living matter from relatively simple molecules thought to have been abundant on the surface of the primitive earth.

The experimental results to date have apparently convinced many scientists that a naturalistic explanation for the origin of life will be found, but there are significant reasons for doubt. In the years since the publication of *Biochemical Predestination* I have been increasingly struck by a peculiar feature of many of the published experiments in the field. I am not referring to those studies conducted more or less along the lines of Miller's original work, although there are firm grounds for criticizing those studies as well. I am referring to those experiments designed to elucidate possible pathways of prebiotic synthesis of certain organic substances of biologic interest, such as purines and pyrimidines, or polypeptides.

In most cases the experimental conditions in such studies have been so artificially simplified as to have virtually no bearing on any actual processes that might have taken place on the primitive earth. For example, if one wishes to find a possible prebiotic mechanism of condensation of free amino acids to polypeptides, it is not likely that sugars or aldehydes would be added to the reaction mixture. And yet, how likely is it that amino acids (or any other presumed precursor substance) occurred anywhere on the primitive earth free from contamination substances, either in solution or the solid state? The difficulty is that if sugars or aldehydes were also present polypeptides would not form. Instead an *interfering cross-reaction* would occur between amino acids and sugars to give complex, insoluble polymeric material of very dubious relevance to chemical evolution. This problem of *potentially interfering cross-reactions* has been largely neglected in much of the published work on the chemical origins of life. The possible implications of such an omission merit careful study.

Other aspects of origin-of-life research have contributed to my growing uneasiness about the theory of chemical evolution. One of these is the enormous gap between the most complex "protocell" model systems produced in the laboratory and the simplest living cells. Anyone familiar with the ultrastructural and biochemical complexity of the genus *Mycoplasma*, for example, should have serious doubts about the relevance of any of the various laboratory "protocells" to the actual historical origin of cells. In my view, the possibility of closing this gap by laboratory simulation of chemical events likely to have occurred on the primitive earth is extremely remote.

Another intractable problem concerns the spontaneous origin of the optical isomer preferences found universally in living matter (e.g., L-rather than D-amino acids in proteins, D-rather than L-sugars in nucleic acids). After all the prodigious effort that has gone into attempts to solve this great question over the years, we are really no nearer to a solution today than we were thirty years ago.

Finally, in this brief summary of the reasons for my growing doubts that life on earth could have begun spontaneously by purely chemical and physical means, there is the problem of the origin of genetic, i.e., *biologically relevant*, information in biopolymers. No experimental system yet devised has provided the slightest clue as to how biologically meaningful sequences of subunits might have originated in prebiotic polynucleotides or polypeptides. Evidence for some degree of spontaneous sequence ordering has been published, but there is no indication whatsoever that the non-randomness is biologically significant. Until such evidence is forthcoming one certainly cannot claim that the possibility of a naturalistic origin of life has been demonstrated.

In view of these and other vexing problems in origin-of-life research, there has been a need for some years now for a detailed, systematic analysis of all major aspects of the field. It is time to re-examine the foundations of this research in such a way that all the salient lines of criticism are simultaneously kept in view. *The Mystery of Life's Origin* admirably fills this need. The authors have addressed nearly all the problems enumerated above and several other important ones as well. They believe, and I now concur, that there is a fundamental flaw in all current theories of the chemical origins of life. Although the tone of the book is critical, the authors have written it in the positive hope that their analysis will help us find a better theory of origins. Such an approach is, of course, entirely consistent with the manner in which scientific advances have occurred in the past.

One of the uniquely valuable features of the book is its discussion (Chap. 6) of the relative geochemical plausibilities of the various types of simulation experiments reported in the literature. To my knowledge this is the first systematic attempt to devise formal criteria for acceptable degrees of interference by the investigator in such experiments. Another especially helpful feature is the detailed discussion of the implications of thermodynamics (Chaps. 7, 8, and 9) for the origin-of-life problem. This important topic is either omitted entirely or is treated superficially in most other books on the chemical origins of life. The authors might have included a more detailed discussion of the problem of optical isomer preferences, but this deficiency detracts in only a minor way from the overall strength of their argument.

If the author's criticisms are valid, one might ask, why have they not been recognized or stressed by workers in the field? I suspect that part of the answer is that many scientists would hesitate to accept the authors' conclusion that it is fundamentally implausible that unassisted matter and energy organized themselves into living systems. Perhaps these scientists fear that acceptance of this conclusion would open the door to the possibility (or the necessity) of a supernatural origin of life. Faced with this prospect many investigators would prefer to continue in their search for a naturalistic explanation of the origin of life along the lines marked out over the last few decades, in spite of the many serious difficulties of which we are now aware. Perhaps the fallacy of scientism is more widespread than we like to think.

One's presuppositions about the origin of life, and especially the assumption that this problem will ultimately yield to a persistent application of current methodology, can certainly influence which lines of evidence and argument one chooses to stress, and which are played down or avoided altogether. What the authors have done is to place before us essentially all the pertinent lines of criticism in one continuous statement and to invite us to face them squarely.

All scientists interested in the origin-of-life problem would do well to study this book carefully and to evaluate their own work in the light of its arguments.

Dean H. Kenyon
Professor of Biology
San Francisco State University

Preface

The Mystery of Life's Origin is a book that had to be written. There is a critical necessity in any developing scientific discipline to subject its ideas to test and to rigorously analyze its experimental procedures. It is an ill-fated science that doesn't do so. Yet, surprisingly, prebiotic or chemical evolution has never before been thoroughly evaluated. This book not only provides a comprehensive critique using established principles of physics and chemistry, it introduces some new analytical tools, particularly in chapters six and eight.

We do not want to suggest that scholars have offered no criticisms helpful to other workers in the field of origin-of-life studies. They have, of course, and scattered here and there in the chemical evolution literature these criticisms can be found. There is no comprehensive marshalling of these, however, no carefully ordered statement that brings them together in one volume to assess their combined import. That is a need that has existed now for several years, a need which, hopefully, this book helps remedy. It should not be thought that the authors cited as sources of specific criticisms would be in agreement with the overall reassessment presented here. In most cases they would not.

The fact that chemical evolution has not received thorough evaluation to date does not mean it is false, only that it is unwise to build on it or extend it until we are satisfied it is sound. It is crucial to have a thorough critique of chemical evolution, expecially since much of the optimism about finding life in space and the search for extraterrestrial intelligence (SETI) is based on it.

Workers who have come up within a discipline usually are the ones most qualified to administer criticism. There are times, however, when workers with specialized training in overlapping disciplines can bring new insights to an area of study, enabling them to make original contributions. The following chapters were produced by a chemist (CT), a materials scientist (WB), and a geochemist (RO). If there is validity to our reassessment it will mean that sizable readjustments in origin-of-life studies are in order. Even if our critique is shown to be deficient and the chemical evolution scenario is vindicated, perhaps the present work will have played a role in goading scientific workers into presenting a clearer and stronger defense in its behalf.

The authors would like to thank the following for permission to quote extracts or reproduce diagrams from their publications, as indicated in the text: Gordon and Breach, Science Publisher, Inc.: ed. Lynn Margulis, *Origins of Life: Proceedings of the First Conference*, 1970; Simon & Schuster, Inc.: F. Hoyle and N.C. Wickramasinghe, *Evolution from Space*, 1981; Marcel Dekker: S. Fox and K. Dose, *Molecular Evolution and the Origin of Life*, revised edition, 1977; Prof. A.E. Wilder Smith, *The Creation of Life*, 1970; the MIT Press: ed. J. Neyman, *The Heritage of Copernicus*, 1974; and the American Chemical Society: S.W. Fox, K. Harada, G. Krampitz, and G. Mueller, *Chemical Eng. News*, June 22, 1970.

It is a pleasure to acknowledge the help, counsel, advice, criticism, and encouragement of many colleagues and friends. Without their assistance it is doubtful this book would have been written. Any errors of facts or interpretation, however, are our own. In particular for reading and commenting on the entire manuscript, we would like to thank Frank Green, Robert L. Herrmann, Dean Kenyon, Gordon Mills, G. Shaw, Grahame Smith, Peter Vibert and John C. Walton. And for expert comments on individual chapters we thank Greg Bahnsen, Art Breyer, Tom Cogdell, Preston Garrison, Norman Geisler, Harry H. Gibson, Jr., Charles Hummel, Glenn Morton, Francis Schaeffer, David Shotton, David Stephens and Hubert Yockey.

Finally, our heart-felt thanks go to our wives, Carole, Ann, and Candace, who endured through seven years of research, writing and revisions. It is to them whose loyalty and love never waned that we dedicate this book.

Dallas, Texas
Christmas, 1983

C. Thaxton
W. Bradley
R. Olsen

Contents

CHAPTER 1

Crisis in the Chemistry of Origins

Two monumental scientific reports appeared in 1953, both of which have subsequently received wide acceptance in the scientific community. One was the proposal by James Watson and Francis Crick[1] of their double helical model for deoxyribosenucleic acid, or DNA. According to their now-famous model, hereditary information is transmitted from one generation to the next by means of a simple code resident in the specific sequence of certain constituents of the DNA molecule. It had previously been held that the spectacular diversity of life was due in part to some corresponding diversity of nuclear material. The breakthrough by Crick and Watson was their discovery of the specific key to life's diversity. It was the extraordinarily complex yet orderly architecture of the DNA molecule. They had discovered that there is in fact a code inscribed in this "coil of life," bringing a major advance in our understanding of life's remarkable structure.

Almost as if synchronized for the sake of irony, the other report in 1953, by Stanley Miller[2], offered experimental support for what has become an increasingly apparent contradiction. Miller offered his work in support of the neo-Darwinian theory of prebiotic evolution. This notion suggested that the fantastic complexity in the molecular organization of living cells might somehow have resulted from nothing more than simple chemicals interacting at random in a primordial ocean.

In 1953, few if any were troubled by the tension between the new insights of Crick and Watson on the one hand and Miller's results on the other. Crick and Watson were concerned with life's *structure* and Miller was concerned with life's *origin*. Most observers had an unshakable confidence that these two investigative approaches would eventually converge. After all, young Miller's announcement of experimental success was just what was anticipated according to the general theory of evolution. Regardless of whether the particular theory of evolution is Darwinian, neo-Darwinian, or something else, an evolutionary preamble to the biological phase of evolution is clearly required. *Chemical evolution*, then, is the pre-biological phase of evolution in which the very earliest living things came into being. This monumental dawning of life occurred through the variation of natural forces acting on matter over long time spans, perhaps up to a thousand million years, or maybe longer.

In the decades since Miller's and Crick and Watson's reports, however, there have been indications that all is not well in the halls of biology. We have gained a far deeper appreciation of the extremely complex macromolecules such as proteins and nucleic acids. The enlarged understanding of these complexities has precipitated new suggestions that the DNA mechanism may be more complex and the molecular organization more intricate and information-filled than was previously thought.[3]

The impressive complexities of proteins, nucleic acids, and other biological molecules are presently developed in nature only in living things. Unless it is assumed such complexity has always been present in an infinitely old universe, there must have been a time in the past when life appeared *de novo* out of lifeless, inert matter. How can the mere interaction of simple chemicals in the primordial ocean have produced life as it is presently understood? That is the question. The signs do not bode well for the standard answers given, and some investigators are suggesting that our two approaches will not converge.

The Demise of the Role of Chance

By 1966 a major change in scientific thought was underway. In Philadelphia a symposium was held to highlight these changes.[4] It was there that signs of an impending crisis first emerged. Symposium participants came together to discuss the neo-Darwinian theory of evolution. One conclusion, expressed in the words of Murray

Eden of MIT, was the need "to relegate the notion of randomness to a minor and non-crucial role"[5] in our theories of origins. This conclusion was based on probability theory, which shows mathematically the odds against the chance formation of the highly complex molecular structure required for life. With the help of high-speed computers, programs could be run which simulated the billions-of-years' process based on the neo-Darwinian model of evolution. The results showed that the complexity of the biochemical world could not have originated by chance even within a time span of ten billion years. Eden's conclusion was a reasonable if unsettling one.

Other symposium participants voiced similar views about chance or randomness. V.F. Weiskopf noted, "There is some suspicion that an essential point [about our theories of origins] is still missing."[6] Eden suggested "new laws" as the missing piece in the puzzle of life's origin.[7] In his opening remarks as chairman, Nobel Prize-winning biologist Sir Peter Medawar said, "There is a pretty widespread sense of dissatisfaction about what has come to be thought of as the accepted evolutionary theory in the English-speaking world, the so-called neo-Darwinian theory."[8] It was Marcel Schutzenberger of the University of Paris, however, who intimated the true extent of the developing crisis when he expressed his belief that the problem of origins *"cannot be bridged within the current conception of biology"*.[9] (Emphasis added).

These comments reflect the impotence of chance or randomness as a creative mechanism for life's origin. But there was dissent, too. Some symposium participants, C.H. Waddington for example, balked at this conclusion, saying that faulty programming was the problem, not chance.[10] Waddington's objection illustrates a basic dilemma that has always plagued probability calculations. Such calculations must first assume a plausible chemical pathway, or course of events, and then calculate the probability of this series of events, in the hopes that the answer will at least approximate the probability of the *actual* course of events. Nevertheless, there is great uncertainty about the actual chemical pathway. As a consequence, calculations showing the extreme improbability that life began by chance usually have carried little weight with scientists.

Such probability calculations, however, have now been supplemented by a more definitive type of calculation which does not require a knowledge of the detailed process or exact path of events that led to life. Recent advances in the application of the first and second laws of thermodynamics to living systems provide the basis for these calculations. Through them, accurate probabilities for the spontaneous synthesis of complex chemicals can be calculated without regard to the path that led to their development. All that is needed is information about the initial chemical arrangement and the complex arrangement these chemicals are found to have in living things. These thermodynamic calculations have agreed in order of magnitude with earlier path-dependent probability calculations. For example, some investigators, including Ilya Prigogine, the Nobel Prize-winning thermodynamicist, have relied upon calculations based on equilibrium thermodynamics to show the probability that life occurred spontaneously. Prigogine et al., put it this way:

> The probability that at ordinary temperatures a macroscopic number of molecules is assembled to give rise to the highly ordered structures and to the coordinated functions characterizing living organisms is vanishingly small. The idea of spontaneous genesis of life in its present form is therefore highly improbable even on the scale of the billions of years during which prebiotic evolution occurred.[11]

The agreement between the two types of probability calculations has heightened the growing awareness of a crisis in the chemistry of origins.

Biochemical Predestination

Because of the increasing disillusionment with the role of chance, a shift took place in the late Sixties and the Seventies to the view that life was somehow the *inevitable* outcome of nature's laws at work over vast spans of time. Terms such as "directed chance" and "biochemical predestination" have entered the scientific literature to mean that life was somehow the result of the inherent properties of matter. The abundant use of these terms marks a shift in thinking. Many feel that bonding properties of atoms had a significant role in the origin of the complex molecular structures of life. Others, including M. Polanyi, however, have suggested that if atomic bonding properties accounted for the actual structure of DNA, including the distribution of bases, "then such a DNA molecule would have no information content. Its codelike character would be effaced by an overwhelming redundancy."[12] So the mystery behind life's origin continues in spite of the undaunted confidence of some that a solution is near. This is further illustrated by developments in the U.S. space program.

In 1974 Stanley Miller, who had continued in his efforts to put modern origin-of-life studies on a firm experimental footing, said that:

> We are confident that the basic process [of chemical evolution] is correct, so confident that it seems inevitable that a similar process has taken place on many other planets in the solar system.... We are sufficiently confident of our ideas about the origin of life that in 1976 a spacecraft will be sent to Mars to land on the surface with the primary purpose of the experiments being a search for living organisms.[13]

In 1976, on the eve of the first Mars landing, NASA's chief biologist, Harold P. Klein, explained that if our theories of origins are correct, we should find corroborative evidence of it on Mars.[14] The theories of which he spoke had presupposed the same inexorable forces of chemistry and physics at work on Venus, Mars, and innumerable planets throughout the cosmos as on earth. Although few space scientists actually expected to find life on Mars, there was wide agreement that organic chemicals in some stage of the life-forming process would likely be found there. And, of course, the cost of the Mars landing was a substantial pledge toward that confidence. A significant opportunity for confirmation had arrived. The origin-of-life experiments were disappointing, however, as an unexpected type of chemistry was found on Mars, which indicated environmental conditions unfavorable to chemical evolution.

In a detailed analysis of the Mars data as reported in the *Journal of Geophysical Research*, it was concluded that "the results of the organic analysis experiment...should not give encouragement to those who hope to find life on Mars."[15]

Results from the Voyager I fly-by of Jupiter and Saturn have not given any additional encouragement to those hoping to discover life in the cosmos other than that on earth.[16] One of Saturn's moons, Titan, was thought to be more hospitable toward life. It now appears that Titan is cold and dead, with an atmosphere about 85% nitrogen, 15% argon, and less than 1% methane.[17]

The sticky question that remains unresolved is not merely whether objections raised at the Wistar Institute are correct, but, in light of current evidence, whether there is cause for optimism about the "directed chance" view of life's origin. What is responsible for the dashed expectations held about life first on the moon, then Venus, Mars, Jupiter, and now Saturn and its moon Titan?

It cannot be denied that the "pure chance" view of the origin of life is a position of extreme faith. The nagging difficulty, however, that faces us now is that the modified version of "directed chance" has not performed well to date. The question must be asked whether there is a flaw in our theory of chemical evolution—a flaw at such a fundamental level that it mars both theories, "pure chance" and "directed chance."

A flaw in our theory of origins need not be viewed with pessimism, however. H.R. Post, a philosopher of science, has suggested in an illuminating article that such a flaw might actually lead to a new even better theory, if we but learn to decipher properly the clues it can yield. Post said:

> The best workers [in scientific theory] are those who are best at noticing cracks, anomalies, in the existing structure of the old theory—not disagreements with (new) experimental data, but anomalies within the theory itself. These cracks are very strong hints that suggest the structure of the new theory: we infer, as it were, the nature of the new three-dimensional beast from its two-dimensional footprints. They are traces of the new theory in the old.[18]

So for now we assert there is a crack in all current theories of origins. We shall leave for the main body of the book the task of mapping out the contours of the crack, which we hope will further a better understanding of origins.

Speculative Reconstructions

Before coming to that, however, it will be valuable to consider how origin-of-life research relates to science as a whole. In the matter of origins, there were no observers present. For some this lack of observation entirely removes the question of life's origin from the domain of legitimate science. In another context, George Gaylord Simpson has observed that:

> It is inherent in any acceptable definition of science that statements that cannot be checked by observation are not really *about* anything—or at the very least they are not science.[19] (Emphasis his.)

It is primarily due to this lack of observational check on theories that science cannot provide any *empirical* knowledge about origins. It can only suggest plausible scenarios in an attempt to reconstruct the events that led to the appearance of life on earth.

The strength of physical science lies in its ability to explain phenomena as well as make predictions based on observable, repeatable phenomena according to known laws. Science is particularly weak in examining unique, nonrepeatable events. Commenting on this inherent limitation of science, *Nature* magazine noted:

> Those who work on the origin of life must necessarily make bricks without very much straw, which goes a long way to explain why this field of study is so often regarded with deep suspicion. Speculation is bound to be rife, and it has also frequently been wild. Some attempts to account for the origin of life on the Earth, however ingenious, have shared much with imaginative literature and little with theoretical inference of the kind which can be confronted with observational evidence of some kind or another.[20]

Yes, naturalistic explanations of life's origin are speculative. But does this mean such inquiries are impotent or without value? The same criticism can be made of any attempt to reconstruct unique events in the past. This has not deterred Scotland Yard or the FBI, however, from employing, sometimes with dramatic success, the science of forensic medicine in some bizarre "whodunit." Blood stains and fingerprints are the data of the crime detector and constitute circumstantial evidence in a court of law. Blood stains and fingerprints do not tell their own story, so these data must be fitted into some speculative but plausible scenario to reconstruct what occurred in the past. This kind of scenario is nonetheless only speculation, and no matter how plausible it may be, the truth behind the blood stains and fingerprints may be entirely different from the story alleged. For this reason there is always an element of risk or uncertainty when a jury brings a conviction for a crime based on circumstantial evidence. Juries do bring convictions in such cases, however, and all that is required is that the case be made beyond reasonable doubt. Herein lies the value of the speculative reconstruction of some past event. Although such a speculative scenario may elicit a confession from the defendant, or lead to newly discovered eyewitnesses, its principal value comes from its use as a tool in the hands of a skillful lawyer to make a convincing appeal to the jury which must finally decide the matter.

The study of chemical evolution is strikingly similar to forensic science. Consistent with the uniformitarian view that life arose through processes still going on, numerous investigators have reported on laboratory observations and experiments which they offer as circumstantial evidence for the naturalistic origin of life. Though the conditions of the early earth are assumed to have been different from today's conditions, the processes are assumed to have been the same. According to this uniformitarian thinking, if we can reproduce in our laboratories today conditions as they were in the remote past, we should expect to obtain the kinds of changes that occurred then. This is the basis of *prebiotic simulation experiments* reported in chemical evolution literature.

"Implicit in this [uniformitarian] assumption is the requirement that no supernatural agency 'entered nature' at the time of the origin, was crucial to it, and then withdrew from history."[21] (Actually all that is required for this assumption is that no intelligent—purposive—interruption or manipulation of the workings of natural forces ever occurred at the time of life's origin or since.) The developers of chemical evolution theory acknowledge its speculative nature, but offer it as a highly plausible scenario for the origin of life. We agree that there is scientific value in the pursuit of such reconstructions that should not be dismissed out of hand.

Furthermore, the *source* of our initial assumptions is of little import. It is perfectly legitimate to derive our ideas about what conditions might have been like on the early earth from backward inference from present conditions, intuition, or even from a religious holy book. The scientific criterion is whether this speculative scenario fits the data available and is plausible. Here some clarification is in order. In the familiar Popper[22] sense of what science is, a theory is deemed scientific if it can be checked or tested by experiment against observable, repeatable phenomena. On this basis, relativity theory, atomic theory, quantum theory, germ theory—all have been judged scientific. Since all these theories of science deal with various facets of the *operation* of the universe, let us call them operation theories of science. Our point of clarification notes the difference between operation theories and origin theories, such as theories about the origin of life. Although the various speculative origin scenarios may be tested against data collected in laboratory experiments, these models cannot be tested against the actual event in question, i.e., the origin. Such scenarios, then, must ever remain speculation, not knowledge. There is simply no way to know whether the results from these experiments tell anything about the way life itself originated. In a strict sense, these speculative reconstructions are not falsifiable; they may only be judged plausible or implausible. In fact, as with the speculative scenarios used in a courtroom, failure to render a scenario implausible lends support to its plausibility, its credibility, and enhances the possibility that the reconstruction has genuine explanatory value and is true.

This book is largely devoted to evaluating the speculative scenarios of chemical evolution in light of present and pertinent data. We will seek accurate readings on the progress of various investigative approaches. To set the stage, Chapter 2 will be devoted to a short account of the history and status of chemical evolution theory. Chapter 3 will review representative experiments to simulate chemical events at the monomer level. Chapter 4 begins the critique and main part of the book.

It is our opinion that modern chemical evolution theories of the origin of life are in a state of crisis. The reader will be in a better position to appreciate why we say this after having read the book. But be forewarned! If we are even partially correct, some notable changes are in store for chemical evolution theories. And if we are proven substantially correct, well...but for the time being let's pursue the topic at hand.

References

1. J.D. Watson and F.H. Crick, 1953. *Nature* **171**, 137.
2. S.L. Miller, 1953. *Science* **117**, 528.
3. Anon. 1977. *Nature* **265**, 685; F. Sanger, et al., 1977. *Nature* **265**, 687.
4. P.S. Moorhead and M.M. Kaplan, eds., 1967. *Mathematical Challenges to the Neo-Darwinian Interpretation of Evolution*. Philadelphia: Wistar Institute.
5. Murray Eden, Nov. 1967. In "Heresy in the Halls of Biology," *Scientific Research* p. 59.
6. V.F. Weisskopf in *Mathematical Challenges*, p. 100.
7. Murray Eden in *Mathematical Challenges*, p. 109.
8. Peter Medawar in *Mathematical Challenges*, p. XI.
9. Marcel P. Schutzenberger in *Mathematical Challenges*, p. 73.
10. C.H. Waddinton in response to Schutzenberger's article in *Mathematical Challenges*, p. 73.
11. Ilya Prigogine, G. Nicolis, and A. Babloyantz, Nov. 1972. *Physics Today*, p. 23-31.
12. M. Polanyi, 1968. *Science* **160**, 1309.
13. S.L. Miller, 1974. *The Heritage of Copernicus*, ed. Jerzy Neyman. Cambridge, Mass.: The MIT Press, p. 228.
14. H.P. Klein, July 30, 1976. *The New York Times*.
15. K. Biemann, J. Oro, P. Toulmin III, L.E. Orgel, A.O. Nier, D.M. Anderson, P.G. Simmonds, D. Flory, A.V. Diaz, D.R. Rushneck, J.E. Biller, and A.L. Lafleur, 1977. *J. Geophys. Res.* **82**, 4641.
16. D. Goldsmith and T. Owen, 1980. *The Search for Life in the Universe*. Menlo Park, Calif.: Benjamin/Cummings, p. 106.
17. Geoffrey Briggs and Fredric Taylor, 1982. *The Cambridge Photographic Atlas of the Planets*. Cambridge, Mass.: Cambridge University Press, p. 219.

18. H.R. Post, Feb. 10, 1966. *The Listener* 197.
19. G.G. Simpson, 1964. *Science* **143**, 769.
20. Anon, 1967. *Nature* **216**, 635.
21. D.H. Kenyon and G. Steinman, 1969. *Biochemical Predestination*. New York: McGraw-Hill, p. 30.
22. Karl Popper, 1963. *Conjecture and Refutations*. New York: Harper Torchbooks.

CHAPTER 2

The Theory of Biochemical Evolution

Spontaneous generation has never enjoyed security in prevailing scientific thought. The theory has been alternately embraced, abandoned, and accepted but ignored. The principal reason is that at various times in history two quite distinct concepts have been termed "spontaneous generation." These are: (1) *abiogenesis*, the notion of life's first origin from inorganic matter, and (2) *heterogenesis*, the idea of life's arising from dead organic matter, such as the appearance of maggots from decaying meat.*

The concept of heterogenesis was the more conspicuous of the two, with its apparent observational basis. It was also the more important concept for early Western thinkers. Their Christianized world view seemed to answer the question of life's first origin. Moreover, the vitalistic notion that saw a dichotomy between organic and inorganic matter clearly ruled out the idea of abiogenesis. A long line of Western thinkers, however, including Newton, Harvey, Descartes, and von Helmont, accepted the occurrence of heterogenesis without question.

---

*For additional discussion of the history of spontaneous generation, see "The Spontaneous Generation Controversy (1859-1880): British and German Reactions to the Problem of Abiogenesis," John Farley, 1972. *Journal of the History of Biology*, vol. 5, no. 2, pp. 285-319, from which this discussion has drawn substantially.

The intrigue of the story is that heterogenesis, on the surface a more facile speculation than life from brute chemistry, was put to rest almost simultaneously with the publication of *Origin of Species*. Francesco Redi had demonstrated that meat placed under a screen of muslin (so that flies could not lay their eggs) never developed maggots. Similarly, other examples of heterogenesis were systematically discredited. Schulze, Schwann, von Dusch, and Schroeder each contributed to the growing awareness that microscopic organisms were present in various organic substances.

It was the work of Louis Pasteur, however, which sounded the death knell of the theory of heterogenesis. He showed that air contains many microorganisms which can collect and multiply in water, giving the illusion of spontaneous generation. In 1864, Pasteur announced his results before the science faculty at the Sorbonne in Paris with the words "Never will the doctrine of spontaneous generation recover from the mortal blow of this simple experiment."[1]

The Emergence of Abiogenesis

But the sound of Pasteur's words had not yet stilled before some rcognized that, if taken to its conclusion, Darwin's work required an even more difficult and remarkable form of spontaneous generation—abiogenesis. Even Darwin himself speculated in this regard. In 1871 he wrote in a letter:

> It is often said that all the conditions for the first production of a living organism are now present which could ever have been present. But if (and oh! what a big if!) we could conceive in some warm little pond, with all sorts of ammonia and phosphoric salts, light, heat, electricity, etc. present, that a protein compound was chemically formed ready to undergo still more complex changes, at the present day such matter would be instantly devoured or absorbed, which would not have been the case before living creatures were formed.[2]

The breakdown of the dichotomy of organic and inorganic matter had by this time occurred. The primary impetus in its demise was the Reductionist school of thought which maintained that living matter had no autonomous vital forces within. The reductionist school had drawn support from two important breakthroughs, one in the understanding of matter, the other in the understanding of energy. The first came with the synthesis of urea in 1828 by Wohler; this being the first of a variety of organic materials to be synthesized. It is evident that the assumed categorical barrier between the inorganic and organic worlds would be invalidated by such experimental results. The second important occurrence in the turn toward reductionism was the development of the concept of the conservation of energy. If all energies in a reaction can be quantified with no remainder, then no vital force (which had been held to be a kind of energy) was required in the reaction. With these advances for the reductionist viewpoint, a major hurdle had been cleared for the concept of abiogenesis.

In Germany, the quest for a monistic world view (*Weltanschauung*), a consistent and comprehensive philosophical explanation of all things, demanded a lively debate about abiogenesis. Ernst Haeckel, the most influential of the German evolutionary monists in the two decades following the publication of *Origin of Species* sought earnestly to ensure that the *Weltanschauung* was built around evolution. The dogmatic materialists added their zeal to the same effort.

In contrast, scientists in Britain refused to enter the discussion, at least for a time. British scientists not only resisted the ideas of the monists, but regarded themselves in the traditions of Newtonian science and J.S. Mill. The London *Times* captured their spirit well when it said, "We look to men of science rather for observation than for imagination."[3] World views to a British scientist were apt to be regarded as grandiose speculations, unbecoming to science.

By the 1870s, however, the rigidity of this approach was somewhat mitigated, and Henry Bastian, heavily influenced by Haeckel, argued for a continuous abiogenesis. Bastian saw protoplasm as a simple, undifferentiated substance, arising over relatively brief periods of time on many occasions. Huxley had linked biological evolution and the geological principle of uniformity, and Bastian's view seemed to make sense in that light. We should recognize that at the time, the earth's atmosphere was considered to have been the same in the distant past as in the present. Bastian's notion seemed to be consistent with the principle of uniformity, which gave it added status to many. Indeed, by calling on the reductionist continuity of organic and inorganic matter, Bastian argued that evidence for heterogenesis (still lingering in his own experiments) was evidence for abiogenesis as well. Thus, until the discovery of heat resistant spores, it appeared that Bastian could actually offer experimental support for a continuous abiogenesis. But with the discovery of such spores, the case for abiogenesis was reduced to the argument of a few proponents.

By 1880, not only experiment but even most of the discussion about abiogenesis was abandoned. While subsequent thinkers were to speculate that living matter had greater complexity than Bastian's conception, it was not until the elemental nature of matter was understood that a modern theory of abiogenesis could be forged.

Then is 1924, after years of virtual silence, the Russian biochemist Alexander Ivanovich Oparin reopened the discussion by proposing that the complex molecular arrangements and functions of living systems evolved from simpler molecules that preexisted on the lifeless, primitive earth.[4] With this suggestion, the recognizably modern form of chemical evolution theory began to develop.

In 1928, the British biologist J.B.S. Haldane published a paper in the *Rationalist Annual* in which he speculated on the early conditions necessary for the emergence of terrestrial life.[5] Haldane pictured ultraviolet light acting upon the earth's primitive atmosphere as the source of an increasing concentration of sugars and amino acids in the ocean. He believed that life eventually emerged from this primordial broth. Later, work by J.D. Bernal in 1947 elaborated. Bernal suggested some possible mechanisms whereby biomonomers might accrue to concentrations sufficient to allow condensation reactions producing the macromolecules necessary for life.[6] Both marine and fresh-water clays were seen as instrumental in the synthesis of large macromolecules, as well as their protection from destruction by ultraviolet light.

A further critical contribution to the idea was made by Harold Urey. Urey observed that with the exception of the earth and the minor planets, the solar system was reducing, being hydrogen rich in all the planetary atmospheres. Urey suggested that perhaps the early earth's atmosphere was reducing as well, and that only later in the earth's evolution did it become an oxidizing atmosphere.[7] This concept provided for favorable conditions for the synthesis of organic compounds.

The Modern Theory of Chemical Evolution

The foundational suggestions of Oparin, Haldane, Bernal, and Urey have since been elaborated into what we shall call the modern theory of chemical evolution. This theory came to predominate the thinking of scientists in the latter half of this century. A well-established central core has become the basis for many variations as the theory has developed. In outline form, the general scheme is quite simple. It envisions that the atmosphere of the early earth contained such gases as hydrogen, methane, carbon monoxide, carbon dioxide, ammonia, and nitrogen, but no free oxygen. While this atmosphere would be quite toxic to us, its reducing quality was hospitable to organic molecules. This atmosphere is the first of five stages in the schematic representation of chemical evolution shown in fig. 2-1.

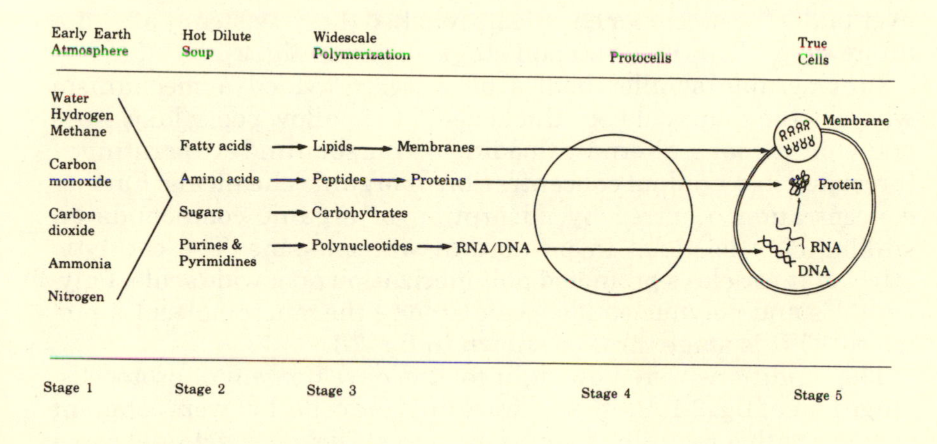

Figure 2-1.
Major stages of chemical evolution.

Sometime close to 3.5 billion years ago, the earth's surface had cooled to under 100°C. This allowed for the survival of various organic molecules that would have degraded in higher temperatures.

Various forms of energy bathed the primitive earth. These energy sources—lightning, geothermal heat, shock waves, ultraviolet light from the sun, and others—drove reactions in the atmosphere and ocean to form a wide variety of simple organic molecules. In the upper zones of this primitive atmosphere there would have been little, if any, free oxygen with which ultraviolet light could interact to produce an ozone layer such as presently protects all living things from lethal doses of ultraviolet. Instead, ultraviolet would irradiate the reducing atmosphere to form amino acids, formaldehyde, hydrogen cyanide, and many other compounds.

At lower altitudes these same organic compounds would result from the energy in electrical storms and thunder shock waves. Synthesis would be occasioned at the earth's surface by wind blowing gases of the reduced atmosphere over hot lava flows near the sea.

The simple compounds formed in the atmosphere were washed down by rain into the oceans. Here they gathered with the products of ocean reactions as abundant organic material began to accumulate. Further reactions inevitably took place in this reservoir, and eventually the precursor chemicals reached the consistency of a "hot dilute soup." This is the second stage shown in fig. 2-1.

Innumerable smaller bodies of water provided a mechanism whereby the soup could be "thickened." In shallow pools, lakes, and shoreline lagoons, alternate flooding by the soup and evaporating of it resulted in a gradual concentration of organic chemicals. Further concentration occurred by adsorption of organic compounds on sinking clay particles in primordial water basins. The catalytic effect of these clays promoted polymerization on a wide scale. Polypeptides and polynucleotides were among the macromolecules produced. This is stage three as shown in fig. 2-1.

The conditions were now right for the development of protocells, stage four of fig. 2-1. Protocells were not true cells, but were coherent systems with a retaining membrane and sufficient functional capacity to survive an interim period. Over this period of time, their internal complexity increased. Polypeptides with suitable specificity to become enzymes developed. Additional characteristics of living cells emerged. When the nucleic acids—life's hereditary molecules—became sufficiently developed, they took control of these processes. Finally, life itself gained its critical first foothold, stage five of fig. 2-1.

This general outline has provided a rich basis for extensive study and numerous laboratory experiments. The theory maintains that natural processes alone operated to form life on this planet. No mysterious, divine, or vital forces had a part. As Cyril Ponnamperuma put it, "...life is only a special and complicated property of matter, and...*au fond* [basically] there is no difference between a living organism and lifeless matter..."[8] The question scientists have pursued, however, is, exactly what were the natural forces?

The neo-Darwinian view is mechanistic in nature. It has seen *extrinsic* forces bringing the increasing order as a result of their chance operation upon the chemical compounds involved. The materialistic view, on the other hand, is the view that matter's *intrinsic* properties are somehow responsible for its own increasing complexity. Life is seen as the inevitable result of the outworking of these intrinsic properties. This view gradually gained ascendancy in the Seventies. Whether called "biochemical predestination" or some other name, it came to enjoy new prestige in the theoretical shift highlighted at the Wistar Institute, as mentioned in Chapter 1.

References

1. R. Vallery-Radot, 1920. *The Life of Pasteur*, from the French by Mrs. R.L. Devonshire. Doubleday, New York: p. 109.
2. Francis Darwin, 1887. *The Life and Letters of Charles Darwin*. New York: D. Appleton, II, 202. Letter written 1871.
3. London *Times*, Sept. 19, 1870.
4. A.I. Oparin, 1924. *Proiskhozhdenie Zhizni*, Izd. Moskovski Rabochii, Moscow. *Origin of Life*, trans. S. Morgulis. New York: Macmillan, 1938.
5. J.B.S. Haldane, 1928. *Rationalist Annual* **148**, 3-10.
6. J.D. Bernal, 1949. "The Physical Basis of Life," paper presented before British Physical Society; *The Physical Basis of Life*. London: Routledge, 1951.
7. H.C. Urey, 1952. *The Planets: Their Origin and Development*. New Haven: Yale University Press.
8. Cyril Ponnamperuma, 1964. *Nature* **201**, 337.

CHAPTER 3

Simulation of Prebiotic Monomer Synthesis

We may wish that a crack team of scientific observers had been present to record and detail the origin of life when it occurred. But since there were no observers, and since we can't go back to investigate the primitive earth, we must do what we can to gain after-the-fact evidence of what may have occurred. We certainly can simulate in the laboratory what we postulate were the conditions of the early earth and, using the uniformitarian principle, assume that the results will be similar to what actually occurred on the prebiotic earth. With this expectation before us, the challenge is one of seeking to accurately identify and reproduce conditions of the prebiotic earth for our experiments. Many noteworthy efforts have been made. But as we shall see, mimicking the early earth is tricky business.

How to Run a Prebiotic Simulation Experiment

For example, we could run our simulation experiment simply by trying to reproduce early earth conditions in a huge enclosed vat containing the suspected chemicals. The experiment would be conducted by passing various energy sources through a mixture of simple gases, liquid water, sand, clay, and other minerals, and just letting it go. Then at various times a portion could be withdrawn for analysis and the progress charted. Such a procedure—a "Synthesis in the Whole"—has on occasion been suggested.[1]

There are criticisms of this approach, however. First, if it truly simulated early earth conditions and processes, we should not expect any meaningful results within laboratory time. Millions of years of simulation might be required for any detectable progress. Second, this method would obscure the complex chemical interactions sought for observation by allowing literally thousands of different reactions to go on simultaneously. This points out the need for a method of partitioning or isolating the various chemical reactions. Only through such partitioning can we gain clues as to the mechanisms involved in the production of life. So we would predictably learn nothing of consequence from a "Synthesis in the Whole" approach.

What we need is some technique which allows us to single out individual reaction processes in our simulated "prebiotic soup" and thus follow their progress. Such an approach would allow us to say something meaningful about the mechanism that might have been involved in the pathway to life, and also about the validity of the proposed scheme itself.

In addition, for a laboratory simulation experiment to be of practical value, some technique must be used to overcome the factor of millions of years of time. Somehow we must speed up the process so that, like time-lapse photography, we are able to effectively compress the happenings of a long time span into manageable laboratory time, yet without distortion.

In fact it is widely accepted today that a technique is available for simulating the extended time factor and for charting the progress of individual chemical reactions. The technique consists of carefully selecting and purifying chemicals conceived to have been the probable precursors of life and subjecting them in mixture to geologically plausible conditions of heat, light, temperature, concentration, pH, etc. An experiment is said to be geochemically plausible when the conditions used reproduce to a substantial degree the conditions alleged for the primitive earth. These experiments are deemed successful if biologically significant molecules or their precursors are found among the products.[2]

In this way, an initial experiment can be run to produce amino acids. Then after isolating, purifying, and concentrating them, the next stage can be simulated, reacting the amino acids together to form polymers. After a similar process of isolating, purifying, and concentrating these polypeptides, the next stage could be simulated in a third experiment to see what is produced. By following this procedure, products such as polysaccharides, lipids, polynucleotides, and protocells might all conceivably result. In time it is hoped that through the right experimental conditions in appropriate prebiotic simulation techniques, a living entity will be produced. Such an accomplishment, it is widely regarded, would lend a great deal of support to the view that life occurred on this planet by natural means. In this chapter we will give a representative review of the kind of simulation experiments at the monomer stage that have been done, and their results.

Table 3-1 shows the relative abundances of present sources of energy averaged over the earth. We shall use this as a guide for the availability of energy sources on the early earth. In the experiments discussed, five energy sources will be considered: electrical discharges, heat, ultraviolet light, shock waves, and high-energy compounds. There are a number of comprehensive reviews of prebiotic simulation experiments.[3] Readers are directed to them for more details.

Table 3-1.
Sources of energy on the Contemporary Earth.

Source	Energy (cal cm^{-2} yr^{-1})
Total radiation from sun	260,000
Ultraviolet light	
$\lambda < 3000$ Å	3,400
$\lambda < 2500$ Å	563
$\lambda < 2000$ Å	41
$\lambda < 1500$ Å	1.7
Electric discharges	4
Cosmic rays	0.0015
Radioactivity (to 1.0 km depth)	0.8
Volcanoes	0.13
Shock waves	1.1
Solar wind	0.2

(From S. Miller, H. Urey and J. Oro, 1976. *J. Mol. Evol.* **9**, 59.)

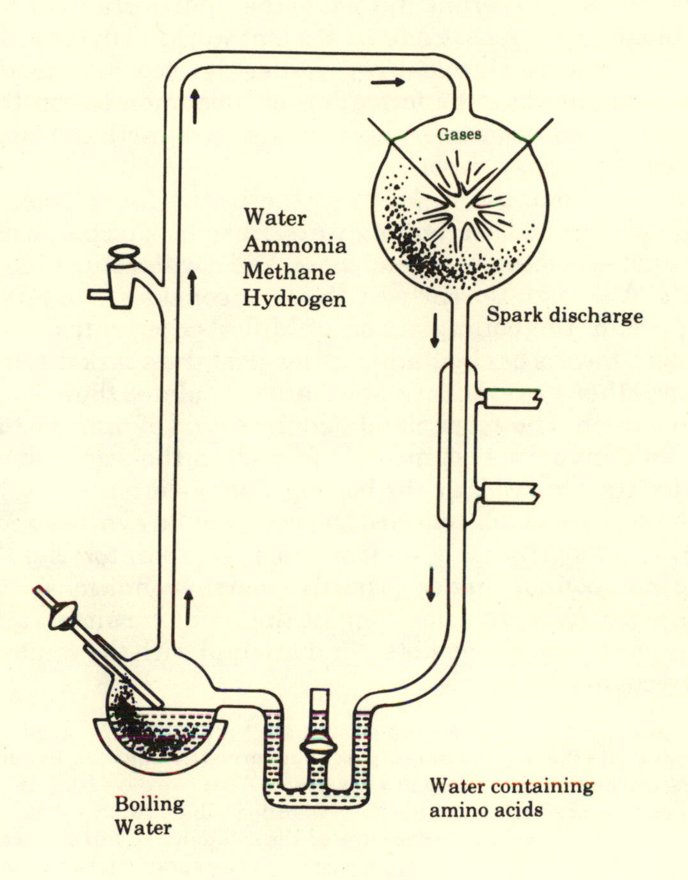

Figure 3-1.
Apparatus used in Miller's electrical discharge experiments to form amino acids. (After R. Jastrow, 1967. *Red Giants and White Dwarfs*. New York: Harper and Row, p. 134.)

Survey of Prebiotic Simulation Experiments

Electrical Discharge Experiments

At the University of Chicago in December 1952, Stanley Miller provided the first experimental test of the Oparin-Haldane hypothesis of abiogenesis.[4] As a graduate student working in the laboratory of Nobel Laureate Harold Urey, Miller devised an experimental approach to simulate the formation of biomonomers on the early earth. The simulated atmosphere consisted of methane, ammonia, hydrogen, and water vapor.

The pyrex apparatus for Miller's experiment (fig. 3-1) consisted of a small boiling flask containing water, a spark discharge chamber with tungsten electrodes, a condenser, and a water trap to collect the products. Although the early earth is not considered to have had a boiling ocean, the boiling action of Miller's apparatus provided a convenient means of circulating gases past the spark discharge.

In most other ways, Miller's apparatus simulated the events on the primitive earth. The spark simulated the action of rain washing into the ocean nonvolatile compounds formed in the atmosphere. And the water trap (as well as the boiling flask) simulated the oceans, pools, and lakes which collected the compounds synthesized.

In 1974 Miller gave an account of "the first laboratory synthesis of organic compounds under primitive earth conditions."[5] In his reminiscence we learn something of the prebiotic simulation technique he used. Describing his second attempt with the same apparatus, he recounts:

> Again after about a week's work getting everything ready, I filled up the apparatus with the same mixture of gases and turned the spark on, keeping the heating coil on the 500-ml flask at a low heat.... After two days I decided to see what had been produced. This time there were no visible hydrocarbons, but the solution was a pale yellow. I concentrated the solution and ran a paper chromatogram. This time I found a small purple spot on spraying with ninhydrin that moved at the same rate as glycine, the simplest amino acid.[6]

As he continues, we pick up some of the drama of those early experiments late in 1952.

> I set the apparatus up again and this time boiled the water more vigorously.... In the morning when I looked at the apparatus the solution looked distinctly pink.... My immediate thought was porphyrins ...and I rushed over to Urey and brought him back to see the color, which he viewed with as much excitement as I did.

> At the end of the week, I removed the solution and did a little processing on it and then ran a two-dimensional paper chromatogram.... This time seven purple spots showed up on spraying with ninhydrin. Three of these amino acids were strong enough and in the correct position to be identified as glycine, α-alanine, and β-alanine.[7]

Since those early days of groundbreaking in the history of simulating prebiotic events, electrical discharge experiments have been repeated many times using a variety of atmospheric compositions. These have included mixtures of two or more of the following gases: methane, ethane, ammonia, nitrogen, water vapor, hydrogen, carbon monoxide, carbon dioxide, and hydrogen sulfide. By and large these experiments follow the same general technique used by Miller, although a number of modifications have been employed. As long as oxygen has been excluded from the mixture, amino acids and other organic compounds have resulted.

In 1974, Miller reported the amino acids he had obtained in electrical discharge experiments.[8] These are listed in table 3-2. In addition,

Table 3-2.
Yields of amino acids obtained from sparking a mixture of CH_4, NH_3, H_2O and H_2.

Compound	Yield (μM)	Compound	Yield (μM)
Glycine	440	α,γ-Diaminobutyric acid	33
Alanine	790	α-Hydroxy-γ-aminobutyric acid	74
α-Amino-n-butyric acid	270	Sarcosine	55
α-Aminoisobutyric acid	30	N-Ethylglycine	30
Valine	19.5	N-Propylglycine	2
Norvaline	61	N-Isopropylglycine	2
Isovaline	5	N-Methylalanine	15
Leucine	11.3	N-Ethylalanine	< 0.2
Isoleucine	4.8	β-Alanine	18.8
Alloisoleucine	5.1	β-Amino-n-butyric acid	0.3
Norleucine	6.0	β-Amino-isobutyric acid	0.3
tert-Leucine	< 0.02	γ-Aminobutyric acid	2.4
Proline	1.5	N-Methyl-β-alanine	5
Aspartic acid	34	N-Ethyl-β-alanine	2
Glutamic acid	7.7	Pipecolic acid	0.05
Serine	5.0	α,β-Diaminopropionic acid	6.4
Threonine	0.8	Isoserine	5.5
Allothreonine	0.8		

(From S. Miller, 1974. *Origins of Life* **5**, 139.)

asparagine,[9] lysine,[10] and phenylalanine[11] have been reported by others but disputed by Miller.[12]

In all, ten of the twenty proteinous amino acids have been positively identified among the products of electrical discharge experiments, as well as about thirty non-proteinous amino acids. Both tert-leucine and N-ethylalanine have been reported but not definitely confirmed. When more than trace amounts of ammonia have been used, iminodiacetic acid and iminoaceticpropionic acid have resulted. When hydrogen sulfide is added to the gaseous mixture methionine is formed.

In 1963, it was found that a gaseous mixture of methane, ammonia, water vapor, and hydrogen irradiated by an electron beam yielded the heterocyclic base, adenine.[13] In 1983, however, C. Ponnamperuma reported that all five nucleic acid bases found in DNA and RNA have been formed in a single simulated primitive atmosphere experiment.*

In addition, the Miller experiment has shown that formaldehyde and "possibly" some sugars are produced.[14] Experiments by Ponnamperuma have shown that both ribose and deoxyribose can be produced during electron irradiation of methane, ammonia, and water.[15] Table 3-3 shows the relative abundance of the various organic compounds produced in electrical discharge simulation. Note that much more has been done in synthesizing amino acids than other biologically significant molecules, which reflects the relative ease of their production compared to the production of heterocyclic bases, sugars, etc.

Now that many different experiments have been evaluated by scientists, it is widely acknowledged that spark discharge is the most efficient energy source for making HCN and amino acids. However, sparks have been used in laboratory experiments primarily for their convenience. But results to date suggest that spark discharge would not have been an effective energy source for the synthesis of pyrimidines and aldehydes (especially sugars) on the early earth.

Heat Experiments

The heat energy produced today by volcanic activity is about an order of magnitude less than the energy produced by all electrical

* Reported at the 186th National Meeting of the American Chemical Society, August 29, 1983, held in Washington, D.C. See *Chem. Eng. News*, Sept. 5, 1983, p. 4.

Table 3-3.
Yields of organic compounds obtained from sparking a mixture of CH_4, NH_3, H_2O and H_2.

Compound	Relative Yield*
Formic acid	1000
Glycine	270
Glycolic acid	240
Alanine	146
Lactic acid	133
β-Alanine	64
Acetic acid	64
Propionic acid	56
Iminodiacetic acid	24
Sarcosine	21
α-Amino-n-butyric acid	21
α-Hydroxybutyric acid	21
Succinic acid	17
Urea	9
Iminoaceticpropionic acid	6
N-Methyl urea	6
N-Methylalanine	4
Glutamic acid	3
Aspartic acid	2
α-Aminoisobutyric acid	0.4

(After S. Miller, 1974. *Origins of Life* **5**, 139.) Biologically relevant amino acids are written in italics.
*Yields are relative to formic acid and presented in descending order.

discharges (table 3-1) and about the same amount of energy as that produced by lightning. Consequently, a number of workers, the most famous being Sidney Fox, have devised laboratory techniques to simulate "the flow of volcanic gases through fissures or 'pipes' of hot igneous rocks of lava."[16] These experiments are known as thermal synthesis or pyrosynthesis.

The apparatus used in these heat experiments is a modification of the spark apparatus used by Miller. The principal difference is that the spark electrodes have been replaced by a furnace (fig. 3-2). Various "primitive atmosphere" gases are allowed to flow over solid silica gel, alumina, or quartz sand in a furnace kept at 900-1100°C.

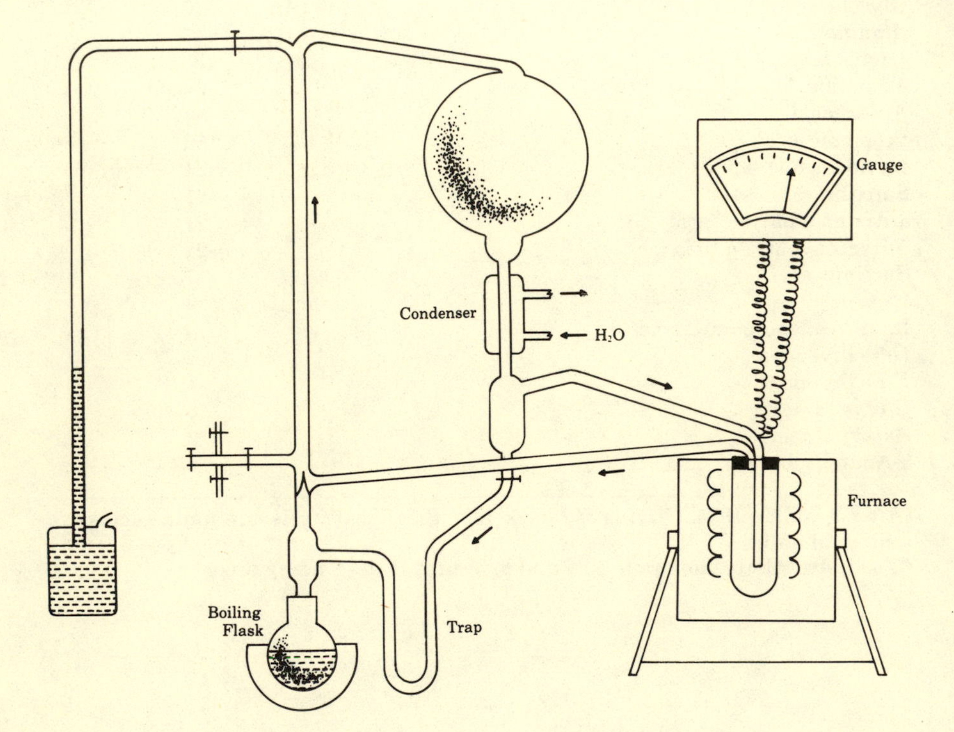

Figure 3-2.
Apparatus used for the thermal synthesis of amino acids from simple gases. (After Harada and Fox, 1965 in *Origins of Prebiotic Systems*. New York: Academic Press, p. 187.)

Customarily, the gases remain in the hot zone for only a fraction of a second, and are then cooled quickly. The products are collected in the trap and then flow into the boiling flask. Table 3-4 shows the results of heating methane, ammonia, and water at 950°C using quartz sand catalyst. Note that twelve proteinous amino acids were reported as dominant products in this experiment by Harada and Fox[17] in 1964. These scientists accounted for the large number of biological amino acids found in terms of a gratuitous role played by heat as an energy source. "According to these [thermal synthesis] results, the contents of unnatural amino acids are depressed and the contents of the natural amino acids enhanced by the use of thermal energy."[18] In addition, four other amino acids found in proteins have subsequently been reported by this heating technique: lysine, tryptophan,

Table 3-4.
Harada and Fox results of heating CH_4, NH_3 and H_2O at 950°C in the presence of quartz sand catalyst.

Amino Acid*	Percent Yield
Aspartic acid	3.4
Threonine	0.9
Serine	2.0
Glutamic acid	4.8
Proline	2.3
Glycine	60.3
Alanine	18.0
Valine	2.3
Alloisoleucine	0.3
Isoleucine	1.1
Leucine	2.4
Tyrosine	0.8
Phenylalanine	0.8
α-Aminobutyric acid	0.6
β-Alanine	?
Sarcosine	
N-Methylalanine	

(From K. Harada and S. Fox, 1964. *Nature* **201**, 335.) Biologically relevant amino acids are written in italics.
*Basic amino acids were not fully studied, and therefore were not listed. Yield is based on percent of total amino acid product.

histidine, and arginine. Efforts have been made to produce the sulfur-bearing amino acids methionine and cysteine by adding hydrogen sulfide. But so far, these attempts have failed.

The reported results of thermal synthesis of amino acids from a simulated primitive atmosphere have been challenged. Lawless and Boynton[19] repeated the experimental procedure described by Harada and Fox, and identified the products by more sophisticated means. As table 3-5 shows, only six amino acids were unequivocally identified, of which only glycine, alanine, and aspartic acid were types found in proteins. It is significant that Fox himself now regards low temperature (i.e. $< 120°C$) routes to amino acids "as the most plausible."[20]

One important variation of thermal syntheses has been the Fischer-Tropsch type technique.[21] In a typical synthesis, carbon monoxide, hydrogen, and ammonia flow through a vycor tube filled with metal or clay catalysts. When heated to 500-700°C for about 1.2

Table 3-5.
Lawless and Boynton results of heating CH_4, NH_3 and H_2O at various temperatures using quartz sand catalyst.

Compound*	Percent Yield+		
	1060°C	980°	930°
Alanine	1	12	4
Glycine	1	59	96
β-Alanine	90	28	
N-Methyl-β-alanine	1.5	1	
Succinic acid	1.5		
β-Amino-n-butyric	1		
Aspartic acid	3		

(From Lawless and Boynton, 1973. *Nature* **243**, 450.) Biologically relevant amino acids are written in italics.

*Compounds identified by gas chromatography and gas chromatography combined with mass spectrometry.

+Yields were determined by amino acid analyzer and gas chromatographic response.

min., the residence time in the tube, they react to yield a variety of amino acids. The usual Fischer-Tropsch synthesis is used industrially to make hydrocarbons from carbon monoxide and hydrogen.

Another version, a "no-flow" or static synthesis, consist of simply heating the gases in a vycor flask at 200-1000°C for 15-16 min., followed by sustained heating at lower temperatures (50-100°C. 15-183 hrs.).

Proteinous amino acids definitely confirmed* in Fischer-Tropsch type syntheses include glycine, alanine, aspartic acid, glutamic acid, tyrosine, lysine, histidine, and arginine.

Ultraviolet Experiments

As pointed out earlier, solar ultraviolet radiation is considered to have been a major energy source on the primitive earth (see table 3-1). Accordingly, some investigators have sought to use ultraviolet radiation in their simulation experiments. However, the major candidates for constituents of the primitive atmosphere (CH_4, CO, N_2, CO_2, H_2S, NH_3, H_2O, H_2) absorb sunlight almost exclusively at wavelengths below 2000 Å. Yet only a minor fraction (0.015%)[22] of incident solar energy occurs at wavelengths this short.[23] Since these constituents absorb only trivial amounts of energy in the necessary wavelengths, little photochemical reaction occurs. However, this is conceptually not a serious limitation. There would have been many millions of years for the small amount of energy available from sunlight to have had its cumulative effect.

In laboratory simulation experiments the simple "primitive" gases are subjected to short wavelength ultraviolet ($<$ 2000 Å) which is derived from the resonance lines of high-intensity emission sources. The simulation apparatus employed is similar to the electrical

---

*There is no generally acceptable criterion for judging the term "definitely confirmed," which is especially a problem for judging published reports prior to about 1970. In the early period often a single analytical technique, e.g., paper chromatography, served to "identify" a particular compound. With improved techniques, thanks largely to space-age developments, it is becoming widely recognized that the appropriate and reliable method of identification of amino acids is analysis by combining gas chromatography with mass spectrometry. Also the traditional approach of determining melting points of the amino acids is reliable, as is the mixed melting point of a suitable derivative. In most experiments, however, not enough material is available for this method.

discharge apparatus used by Miller. The principal difference is that the ultraviolet source replaces the electrodes (fig. 3-3). Results of three such experiments are given in table 3-6 showing noteworthy products.

Table 3-6.
Summary of various simulated atmosphere experiments using ultraviolet light as the energy source to produce amino acids.

Workers	Reactants	Wavelength	Products
Groth and v. Weyssenhoff[24]	Methane, ethane, ammonia, and water vapor	1296 Å and 1470 Å; 1165 Å and 1235 Å	Glycine, alanine, α-amino-butyric acid
Terenin[25]	Methane, ammonia, and water	Continuous UV spectrum	Alanine
Dodonova and Sidorova[26]	Methane, carbon monoxide, ammonia, and water	1450 Å-1800 Å	Glycine, alanine, valine, and norleucine; methylamine, ethylamine, hydrazine, urea, and formaldehyde

In addition to these amino acids, Ponnamperuma has shown that ribose and deoxyribose are produced during ultraviolet irradiation of formaldehyde.[27]

Ultraviolet light would have been the most abundant energy source for the primitive earth (table 3-1). In simulation experiments, however, it has generally given low yields of amino acids. This has usually been interpreted as related to the fact that ultraviolet is not a good source for HCN, a principal intermediate to amino acids through the Strecker synthesis (see below). Ultraviolet light, however, may be the best source for aldehydes, which are also essential intermediates to amino acids by the Strecker mechanism. These results support the widely held belief that a variety of energy sources was responsible for the buildup of concentrations of essential biological precursor chemicals in the primitive oceans.

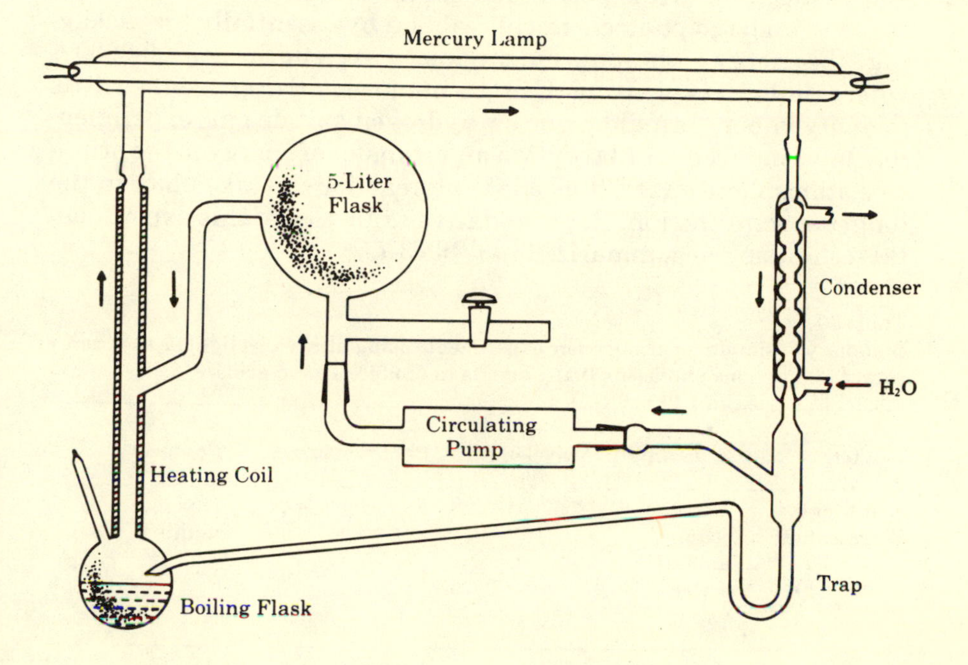

Figure 3-3.
Apparatus used for the mercury-sensitized ultraviolet synthesis of amino acids. (After Kenyon and Steinman, 1969. *Biochemical Predestination*. New York: McGraw-Hill, p. 135.)

Photosensitization

Photosensitization provides a means of overcoming the scarcity of usable ultraviolet light in the early atmosphere. Through this ingenious technique it is possible to get the "primitive" atmospheric gases to undergo photochemical reaction by essentially "repackaging" the energy of the longer ultraviolet wavelengths of 2000-3000 Å where sunlight is plentiful. Using a photosensitizing agent such as mercury vapor, formaldehyde, or hydrogen sulfide gas, experimenters have induced the absorption and transfer of energy to the primitive atmosphere gases, thus enabling reactions to take place in the longer spectral region. Representative examples of this experimental technique are summarized in table 3-7.

Table 3-7.
Summary of simulated atmosphere experiments using ultraviolet light as the energy source and various photosensitizing agents to produce amino acids.

Workers	Reactants	Wavelength	Photosensitizer	Products
Groth and v. Weyssenhoff[28]	Methane, ethane, ammonia, water vapor	2537 Å	Mercury vapor	Glycine, alanine
Sagan and Khare[29]	Methane, ethane, ammonia, water	2537 Å	Hydrogen sulfide	Alanine, glycine, serine, glutamic acid, aspartic acid, cystine
Khare and Sagan[30]			Formaldehyde	
Hong, Hong, and Becker[31]	Ammonia, ethanol	2200-2800 Å Max 2520 Å	Hydrogen sulfide	Serine (or threonine), glycine, alanine, aspartic acid, valine, glutamic acid, leucine, isoleucine, proline

Shock Wave Experiments

According to chemical evolution scenarios, shock waves from thunder and meteorite impact in the atmosphere would have made a small but definite contribution as an energy source on the prebiotic earth. Investigations have shown that shock waves are very efficient in the synthesis of amino acids from the simple gases of methane, ethane, ammonia, and water vapor. This means that although the overall energy contribution from shock waves may have been small (table 3-1) they could have been a major source of these biomonomers on the early earth.[32]

Shock wave synthesis works by subjecting the gases to a high temperature (2000-6000° K) for a small fraction of a second followed by rapid cooling.[33] Thus far this technique has resulted in the following amino acids: glycine, alanine, valine, and leucine.

High-Energy Chemicals

Most of the amino acids found in proteins have been identified as products in experiments using aqueous solutions to simulate the primordial ocean. Although many of these experiments still use heat or ultraviolet light, most do not require an outside energy source. Instead, reactions are found to go spontaneously by the use of high-energy chemicals such as hydrocyanic acid, cyanate, cyanoacetylene, formaldehyde, hydroxylamine, or hydrazine. The warrant for their use in ocean simulations is their presence among the products of atmosphere experiments.

Many of these high-energy compounds would have had double or triple bonded carbon atoms. Common examples would be the ethylenes ($>C=C<$), acetylenes ($-C\equiv C-$), aldehydes ($RCH=O$), ketones ($R_1R_2C=O$), carboxylic acids ($RCOOH$), and nitriles ($RC\equiv N$). These compounds would enter into reactions directly by using the energy released by their double and triple bonds. In general they simply add other chemical constituents to their structures by addition across the double or triple bond. For example, ethylene and acetylene will both add water to their structures.

$$\underset{\text{Ethylene}}{CH_2{=}CH_2} + H_2O \rightleftharpoons \underset{\text{Ethyl alcohol}}{CH_3CH_2OH}$$

$$\underset{\text{Acetylene}}{HC{\equiv}CH} + H_2O \rightleftharpoons \underset{\text{Acetaldehyde}}{CH_3CH{=}O}$$

Addition reactions have usually been held to lead to a build-up of a wide assortment of organic compounds in the early ocean. In turn there would have been interaction among these compounds to produce still more complex chemical constituents. For example, two molecules of acetaldehyde could react in aqueous solution to produce acetic acid and ethyl alcohol, as follows:

$$\underset{\text{Acetaldehyde}}{2\ CH_3CH{=}O} + H_2O \rightleftharpoons \underset{\text{Acetic acid}}{CH_3COOH} + \underset{\text{Ethyl alcohol}}{CH_3CH_2OH}$$

These products could then react to yield ethyl acetate:

$$\underset{\text{Acetic acid}}{CH_3COOH} + \underset{\text{Ethyl alcohol}}{CH_3CH_2OH} \rightleftharpoons \underset{\text{Ethyl acetate}}{CH_3COOCH_2CH_3} + H_2O$$

Addition reactions can be envisioned as playing a major role in the production of amino acids. First, two molecules of formaldehyde could react to give aldehyde:

$$\underset{\text{Formaldehyde}}{2\ HCHO} \rightleftharpoons \underset{\text{Glycolaldehyde}}{CH_2OHCH{=}O}$$

Then, two molecules of glycolaldehyde could react with water to yield glycol and glycolic acid:

$$\underset{\text{Glycolaldehyde}}{2\ CH_2OHCHO} + H_2O \rightleftharpoons \underset{\text{Glycol}}{CH_2OHCH_2OH} + \underset{\text{Glycolic acid}}{CH_2OHCOOH}$$

Finally, glycolic acid could react with ammonia to give glycine:

$$\underset{\text{Glycolic acid}}{CH_2OHCOOH} + NH_3 \rightleftharpoons \underset{\text{Glycine}}{CH_2NH_2COOH} + H_2O$$

It has also been suggested that a major synthetic pathway for the formation of amino acids in the primitive ocean would have been the well-known Strecker synthesis. In this synthesis ammonia would have added to an aldehyde carbonyl group to give an imine.

$$\underset{\text{Aldehyde}}{RCH{=}O} + NH_3 \rightleftharpoons \underset{\text{Imine}}{RCH{=}NH} + H_2O$$

Then hydrogen cyanide (HCN) adds to the imine to form an α-aminonitrile.

$$\underset{\text{Imine}}{R\text{-}C\text{-}H{=}NH} + HCN \rightleftharpoons \underset{\alpha\text{-Aminonitrile}}{RCH(NH_2)C{\equiv}N}$$

Finally the synthesis is completed by the irreversible addition of water to the nitrile to form an α-amino acid.

$$RCH(NH_2)C{\equiv}N + H_2O \longrightarrow \underset{\alpha\text{-Amino acid}}{RCH(NH_2)COOH}$$

This is a general synthesis where the amino acid produced depends on the initial aldehyde. For example, by starting with formaldehyde, acetaldehyde, or glycolaldehyde, the amino acids glycine, alanine, or serine, respectively, are produced. Miller and Orgel have shown that many of the 20 amino acids found in proteins could have been formed by the Strecker pathway.[34]

Examples of successful laboratory systhesis are given in table 3-8. Notice that the experiment done by Matthews and Moser produced no fewer than twelve proteinous amino acids.

Table 3-8.
Summary of simulated ocean experiments using high-energy compounds to yield amino acids.

Workers	High-Energy Compound	Medium	Products
Oro and Kamat[35]	Ammonium cyanide	Alkaline aqueous solution	Alanine, glycine, aspartic acid
Lowe, Rees, and Markham[36]	Ammonium cyanide	Alkaline aqueous solution	Glutamic acid, aspartic acid, threonine, serine, glycine, alanine, isoleucine, leucine
Friedmann, Haverland, and Miller[37]	Hydrogen cyanide and acetone	Aqueous ammonia	Valine
Abelson[38]	Hydrogen cyanide	Aqueous	Glycine, alanine, serine, aspartic acid, glutamic acid
Matthews and Moser[39]	Hydrogen cyanide	Anhydrous liquid ammonia	Lysine, histidine, arginine, aspartic acid, threonine, serine, glutamic acid, glycine, alanine, valine, isoleucine, leucine
Pavolovskaya and Pasynskii[40]	Formaldehyde	Aqueous solution ammonium salts	Serine, glycine, alanine, glutamic acid, valine, phenylalanine, isoleucine (the latter only from ammonium nitrate)

Table 3-8 (cont.)

Workers	High-Energy Compound	Medium	Products
Van Trump and Miller[41]	Hydrocyanic acid, acrolein	Aqueous ammonia, methylated hydrogen sulfide, ammonium chloride	Methionine, glutamic acid
Sanchez, Ferris, and Orgel[42]	Cyanoacetylene, hydrocyanic acid	Aqueous ammonia	Aspartic acid, asparagine
Friedmann, Haverland, and Miller[43]	Hydrocyanic acid, phenylacetylene	Aqueous ammonia, hydrogen sulfide	Phenylalanine
Fox, Windsor;[44] Wolman, Miller, Ibanez, and Oro[45]	Formaldehyde	Aqueous ammonia	Aspartic acid, serine, glutamic acid, proline, glycine, and alanine

All of the five bases have been synthesized in solutions which presumably depict oceans and other bodies of water that might have been found on the primitive earth. Adenine was found after aqueous cyanide solutions were heated at 90° C for several days.[46] Both adenine and guanine have been synthesized by the action of ultraviolet light on dilute solutions of hydrocyanic acid.[47]

Of the pyrimidines, cytosine is produced by heating aqueous cyanoacetylene with cyanate for one day at 100° C, or by allowing it to stand at room temperature for seven days.[48] Uracil is formed by heating a solution of malic acid, urea, and polyphosphoric acid to 130° C for one hour.[49] It has also been formed by heating acrylonitrile with urea to 135° C in aqueous solution.[50] This synthesis of uracil has also been successful when using β-aminopropionitrile or β-aminopropionamide instead of acrylonitrile. In addition, it has been found that thymine can be formed by heating uracil with formaldehyde and hydrazine in aqueous ammonia solution for three days.[51]

It would appear from the foregoing experimental evidence that it is fairly easy to form adenine and possibly the other heterocyclic bases. Since adenine is easiest to form and the most stable, we would expect to find it playing important roles in living systems. That is indeed what we find. Some of the most biologically important molecules in living systems are those which contain adenine: DNA, RNA, ATP, ADP, NAD, NADP, FAD, and coenzyme A.

High energy compounds have also been instrumental in the synthesis of sugars. As early as 1861, it was known that sugars could be produced from formaldehyde in dilute aqueous alkaline solution.[52] Since then the method has yielded many different sugars. Examples include: fructose, cellobiose, xylulose, galactose, mannose, arabinose, ribose, xylose, lyxose, and ribulose. Other organic chemicals such as glycolaldehyde, glyceraldehyde, dihydroxyacetone, and a number of tetroses also have been formed by this method.[53] Deoxyribose was produced when solutions of formaldehyde and acetaldehyde were allowed to react at 50°C or less. The base for these solutions was calcium oxide or ammonia.[54] Ribose also has been produced by refluxing formaldehyde solution over the clay mineral kaolinite (a hydrated aluminum silicate).[55]

Summary

As this review demonstrates, there have been many biomonomers produced in these prebiotic experiments. These impressive achievements have included synthesis of nineteen of the twenty proteinous amino acids, all five heterocyclic bases found in nucleic acids, and several essential sugars including glucose, ribose, and deoxyribose. Other likely constituents of the prebiotic soup have been produced as well. Taken together, this is a substantial body of experimental work, and provides the major source of support for chemical evolution theory. These laboratory results have been the basis for much optimism concerning chemical evolution, and many scientists have been virtually assured that the primitive ocean was full of organic compounds. For example, John Keosian said:

> Backed by all the recent experimental evidence, it is now safe to take for granted the existence of a great variety of organic compounds in prebiological times from which to start reconstructing the origin of the first living things.[56]

In a similar vein, Richard Lemmon remarked:

> This research has made it clear that these compounds would have accumulated on the primitive (prebiotic) Earth—that their formation is the inevitable result of the action of available high energies on the Earth's early atmosphere.[57]

References

1. H.H. Pattee, 1961. *Biophys. J.* **1**, 683; A. Rich, 1970. In *Origins of Life: Proceedings of the First Conference*, ed. Lynn Margulis. (Princeton, New Jersey, May 21-24, 1967.) New York: Gordon and Breach, Science Publishers, Inc; Dean H. Kenyon and Gary Steinman, 1969. *Biochemical Predestination*. New York: McGraw-Hill, p. 36.
2. Kenyon and Steinman, *Biochemical Predestination*, p. 284.
3. R.M. Lemmon, 1970. *Chem. Rev.* **70**, 95; C. Ponnamperuma, 1971. *Quart. Rev. Biophys.* **4**, 77; E. Stephen-Sherwood and J. Oro, 1973. *Space Life Sci.* **4**, 5; N.H. Horowitz and J.S. Hubbard, 1974. *Ann. Rev. Genetics* **8**, 393; M.A. Bodin, 1978. *J. Brit. Interplanetary Soc.* **31**, 129-139, 140-146; M. Calvin, 1969. *Chemical Evolution*. New York: Oxford U. Press; Kenyon and Steinman, *Biochemical Predestination*; S.W. Fox and K. Dose, 1972. *Molecular Evolution and the Origin of Life*. San Francisco: W.H. Freeman; J. Brooks and G. Shaw, 1973. *Origin and Development of Living Systems*. New York: Academic Press; S.L. Miller and L.E. Orgel.*1974 The Origins of Life on the Earth*. Englewood Cliffs, New Jersey: Prentice-Hall; K. Dose, S.W. Fox, G.A. Deborin, and T.E. Pavlovskaya, eds., 1974; *The Origin of Life and Evolutionary Biochemistry*. New York: Plenum Press; Lawrence S. Dillon, 1978. *The Genetic Mechanism and the Origin of Life*. New York: Plenum Press.
4. Stanley L. Miller, 1953. *Science* **117**, 528.
5. Stanley L. Miller, 1974. *The Heritage of Copernicus*, ed. J. Neyman. Cambridge: MIT Press, p. 228.
6. Ibid., p. 235.
7. Ibid., p. 235, 236.
8. Miller and Orgel, *The Origins of Life on the Earth*, p. 84. See also: D. Ring, Y. Wolman, N. Friedmann, and S. Miller, 1972. *Proc. Nat. Acad. Sci. USA* **69**, 765; Y. Wolman, W.J. Haverland, and S.L. Miller, 1972. *Proc. Nat. Acad. Sci. USA* **69**, 809; S. Miller, 1955. *J. Am. Chem. Soc.* **77**, 2351.
9. J. Oro, 1963. *Nature* 197, 862.
10. T.E. Pavlovskaya and A.G. Pasynskii, 1959. In *The Origin of Life on the Earth*, eds. A.I. Oparin, et. al., London: Pergamon, 151.
11. C. Ponnamperuma and J. Flores, 1966. *Amer. Chem. Soc. Abstracts*, Meeting, Sept. 11-16.
12. Miller, *The Heritage of Copernicus*, p. 239.

13. C. Ponnamperuma, R.M. Lemmon, R. Mariner and M. Calvin, 1963. *Proc. Nat. Acad. Sci. USA* **49**, 737; C. Ponnamperuma, 1965. In *The Origins of Prebiological Systems and of their Molecular Matrices*, ed. S.W. Fox. New York: Academic Press, p. 221.
14. S.L. Miller and H.C. Urey, 1959. *Science* **130**, 245.
15. Ponnamperuma, in *The Origins of Prebiological Systems and of their Molecular Matrices*, p. 221.
16. Fox and Dose, *Molecular Evolution and the Origin of Life*, p. 84.
17. K. Harada and S.W. Fox, 1964. *Nature* **201**, 335; also K. Harada and S.W. Fox, in *The Origins of Prebiological Systems and of their Molecular Matrices*, p. 187.
18. Harada and Fox, in *The Origins of Prebiological Systems and of their Molecular Matrices*, p. 192.
19. J.G. Lawless and C.D. Boynton, 1973. *Nature* **243**, 450.
20. S.W. Fox, 1976. *J. Mol. Evol.* **8**, 30.
21. D. Yoshino, R. Hayatsu, and E. Anders, 1971. *Geochim. Cosmochim. Acta.* **35**, 927; R. Hayatsu, M.H. Studier, and E. Anders, 1971. *Geochim. Cosmochim. Acta.* **35**, 939; E. Anders, R. Hayatsu, and M.H. Studier, 1973. *Science* **182**, 781.
22. S. Miller, H. Urey, and J. Oro, 1976. *J. Mol. Evol.* **9**, 59.
23. N.H. Horowitz, F.D. Drake, S.L. Miller, L.E. Orgel, and C. Sagan, 1970. In *Biology and the Future of Man*, ed. P. Handler. New York: Oxford U. Press, p. 163.
24. W. Groth and H.v. Weyssenhoff, 1957. *Naturwissenschaften* **44**, 520; 1960. *Planet. Space Sci.* **2**, 79.
25. A.N. Terenin, in *The Origin of Life on the Earth*, p. 136.
26. N. Dodonova and A.L. Sidorova, 1961. *Biophysics* **6**, 14.
27. Ponnamperuma, in *The Origins of Prebiological Systems and of their Molecular Matrices*, p. 221.
28. W. Groth and H.v. Weyssenhoff, 1960. *Planet. Space Sci.* **2**, 79.
29. C. Sagan and B.N. Khare, 1971. *Science* **173**, 417; B.N. Khare and C. Sagan, 1971. *Nature* **232**, 577.
30. B.N. Khare and C. Sagan, 1973. In *Molecules in the Galactic Environment*, eds. M.A. Gordon and L.E. Snyder. New York: John Wiley, p. 399.
31. K. Hong, J. Hong, and R. Becker, 1974. *Science* **184**, 984.
32. A. Bar-Nun, N. Bar-Nun, S.H. Bauer, and C. Sagan, 1970. *Science* **168**, 470; Sagan and Khare, *Science* 417.
33. A. Bar-Nun and M.E. Teuber, 1972. *Space Life Sciences* **3**, 254; A. Bar-Nun and A. Shaviv, 1975. *Icarus* 24, 197.
34. Miller and Orgel, *The Origins of Life on the Earth*, p. 83-117.
35. J. Oro and S.S. Kamat, 1961. *Nature* 190, 442.
36. C.U. Lowe, M.W. Rees, and R. Markham, 1963. *Nature* **199**, 219.
37. N. Friedmann, W.J. Haverland, and Stanley Miller, 1971. In *Chemical Evolution and the Origin of Life*, eds. R. Buvet and C. Ponnamperuma. Amsterdam: North-Holland, p. 123.
38. P.H. Abelson, 1966. *Proc. Nat. Acad. Sci. USA* **55**, 1365.
39. C.N. Matthews and R.E. Moser, 1967. *Nature* **215**, 1230.
40. Pavlovskaya and Pasynskii, in *The Origin of Life on the Earth*, p. 151.
41. J.E. Van Trump and S.L. Miller, 1972. *Science* **178**, 859.
42. R.A. Sanchez, J.P. Ferris and L.E. Orgel, 1966. *Science* **154**, 784.
43. Friedman, Haverland, and Miller, in *Chemical Evolution and the Origin of Life*, p. 123.
44. S.W. Fox and C.R. Windsor, 1970. *Science* **170**, 984.

45. Y. Wolman, S.L. Miller, J. Ibanez, and J. Oro, 1971. *Science* **174**, 1039.
46. J. Oro and A.P. Kimball, 1961. *Arch. Biochem. Biophys.* **94**, 217; Lowe, Rees, and Markham, *Nature*, p. 219.
47. Ponnamperuma, in *The Origins of Prebiological Systems and of their Molecular Matrices*, p. 221.
48. Sanchez, Ferris and Orgel, *Science*, 784.
49. S.W. Fox and K. Harada, 1961. *Science* **133**, 1923.
50. J. Oro, in *The Origins of Prebiological Systems and of their Molecular Matrices*, p. 137.
51. E. Stephen-Sherwood, J. Oro, and A.P. Kimball, 1971. *Science* **173**, 446.
52. A. Butlerov, 1861. *Annalen* **120**, 296.
53. Kenyon and Steinman, *Biochemical Predestination*, p. 146.
54. J. Oro and A.C. Cox, 1962. *Federation Proc.* **21**, 8; J. Oro in *The Origins of Prebiological Systems and of their Molecular Matrices*, p. 137.
55. N.W. Gabel and C. Ponnamperuma, 1967. *Nature* **216**, 453.
56. John Keosian, 1964. *The Origin of Life*. New York: Reinhold Publishing Co., p. 88.
57. Lemmon, *Chem. Rev.*, 95.

CHAPTER 4

The Myth of The Prebiotic Soup

According to Chapter 3 there is a great deal of experimental support for the early stages of chemical evolution. In contrast to the conclusion usually drawn from these experiments, a credible alternative scenario can be presented which argues strongly against chemical evolution.

Although this chapter is essentially critical, our intent is positive. It is not out of malice that a sample of alleged gold is subjected to the refiner's fire. It is done to test the claim of purity, and to burn off dross that precious metal might shine even brighter. Similarly, any good theory should withstand the fires of criticism and be the better for it. In this spirit, we will look at several kinds of difficulties that have persisted for the chemical evolution theory of life's origin. Our purpose is not only to reveal cracks in present origin theories but also to point in the direction of a better theory, i.e., a theory which is in better accord with the data. In general the critique argues that, in the atmosphere and in the ocean, dilution processes would dominate, making concentrations of essential ingredients too small for chemical evolution rates to be significant. Dilution results from the destruction of organic compounds or diminishing the important chemicals for productive interaction. In this chapter we first survey various types of dilution processes. Then, as an example, we estimate how dilute the oceanic soup could have been in essential amino acids. Finally, we consider various mechanisms suggested as means to concentrate the chemical soup.

A Survey of Dilution Processes

According to the original Oparin-Haldane hypothesis from which arose the modern chemical soup theory of origins, ultraviolet light from the sun bathed the prebiotic earth. Together with other sources of energy (e.g., lightning, thunder shock waves, tidal forces, volcanic heat) it would have been sufficient to drive reactions forward.

Simple gaseous molecules of the primitive atmosphere would react to form intermediates and biomonomers. This would be accomplished through the direct absorption of energy. Energy is seen as the means by which molecules can be organized into more complex arrangements, according to the theory.

But energy alone may not be sufficient to increase the complexity or organization of a system. A bull in a china shop does release a great deal of energy, but the effects are mostly destructive. In fact it can be plausibly argued that the energy effects on the early earth would have been very much like the proverbial bull in a china shop. This predominately destructive feature of unbridled solar energy is the first of the several areas of difficulty for the chemical soup theory of life's origin.

Solar Ultraviolet Destruction of Atmosphere Constituents

Concentrations of some of the most important early atmosphere components would have been diminished by short wavelength (i.e., < 2000 Å) ultraviolet photodissociation. Atmospheric methane would have polymerized and fallen into the ocean as more complicated hydrocarbons,[1] perhaps forming an oil slick 1-10 m deep over the surface of the earth.[2] If this occurred, very small concentrations of methane would predictably have remained in the atmosphere. About 99% of the atmospheric formaldehyde would have been quickly degraded to carbon monoxide and hydrogen by photolysis.[3] Carbon monoxide concentrations in the atmosphere would have been small, however. Carbon monoxide would have been quickly and irreversibly converted to formate in an alkaline ocean.[4] Ammonia photolysis to nitrogen and hydrogen would have occurred very quickly, reducing its atmospheric concentration to so small a value that it could have played *no* important role in chemical evolution.[5] If all the nitrogen in the contemporary atmosphere had existed in the form of ammonia in the early atmosphere it would have been degraded by ultraviolet light in 30,000 years.*[6] If the ammonia surface mixing ratio were on the order of 10^{-5} as Sagan has estimated,[7] then the atmospheric lifetime of ammonia would have been a mere 10 years.[8] It would also have been difficult to maintain substantial levels of hydrogen sulfide in the atmosphere. Hydrogen sulfide would have been photolyzed to free sulfur and hydrogen in no more than 10,000 years.[9] The concentration of hydrogen sulfide in the ocean would have been further attenuated by the formation of metal sulfides with their notoriously low solubilities.[10] The same photodissociation process would have applied to water to yield hydrogen and oxygen. Some recent studies suggest that, through ultraviolet photolysis of water vapor, atmospheric oxygen *did* reach an appreciable fraction of today's concentration in early earth times.[11] Naval Research Laboratory results of ultraviolet experiments aboard Apollo 16 suggested that "solar effects on the earth's water may provide our primary supply of oxygen, and not photosynthesis as is generally believed."[12] The principal author of this research, G.R. Carruthers, has however, declared that this news release was "inaccurate" and that photodissociative processes do not rival plant photosynthesis in the production of oxygen.[13] Nevertheless Carruthers is of the opinion that photodissociation of water may have produced perhaps as much as 1% oxygen gas, versus 21% now, in the primitive atmosphere of the first billion years.

Had the primitive oxygen level been even a thousandth part of the present level, it might have been sufficient for an effective ozone screen to form 3-4 billion years ago.[14] If it did, then effectively all ultraviolet wavelengths less than 3000 Å would have been screened from the earth. Such an ozone screen would have deprived the early atmospheric gases of a major energy source. These short ultraviolet wavelengths are lethal to living organisms but are widely considered to have been essential for the origin of life. The issue of oxygen on the early earth is controversial and very important. If the early earth had strongly oxidizing conditions with molecular oxygen present, then spontaneous chemical evolution was impossible.[15]

---

*This estimate was revised to 10^5-10^6 years because of equilibrium of NH_4^+ and NH_3 dissolved in the ocean. (See J.P. Ferris and D.E. Nicodem in *The Origin of Life and Evolutionary Biochemistry*, 1974. Ed. by K. Dose, S.W. Fox, G.A. Deborim, and T.E. Pavlovskaya. New York: Plenum Press, p. 107.

Destruction of Organic Compounds by Energy

Ultraviolet Light. Methane would absorb 1450 Å solar radiation totally by about 30 km. altitude, even if its concentration in the primitive atmosphere were no greater than it is today.[16] Yet theories of life's origin usually allow a substantial methane concentration in the primitive atmosphere. Consequently, syntheses involving the photolysis of methane must have occurred at high altitudes. Amino acids could have been photoproduced at high altitudes from primitive atmospheric gases. Being produced so high they would require perhaps three years (based on fall-out data) to reach the ocean.[17] During this lengthy transport amino acids and other organic compounds would be exposed to the destructive long-wavelength (i.e., > 2000 Å) ultraviolet radiation.[18] This long-wavelength UV is more intense than the short-wavelength (i.e., < 2000 Å) ultraviolet used in synthesis. It has been estimated that perhaps no more than 3% of the amino acids produced in the upper atmosphere could have survived passage to the ocean.[19] Ultraviolet light would also destroy many organic compounds in the ocean since it would penetrate some tens of meters beneath the ocean surface.[20] Ocean currents periodically would surface even the deep water, thus exposing its organic content, too, to destructive ultraviolet light.

Pringle first raised this objection against the effectiveness of primordial synthesis of organic compounds by ultraviolet light in 1954.[21] It has been remarked on many times and continues to be a major objection.

Thermal Decay in Oceans. Organic compounds would have been subject to thermal degradation in the ocean. Based on the thermal half-lives of various organic soup constituents, Miller and Orgel have shown that chemical evolution could not occur if the ocean were warmer than about 25°C, since important intermediates would be destroyed by heat.[22] It is widely held, however, that the average surface temperature of the early earth would have been some 20°C lower than today. This is due to the astronomical theory which says that only about 60% of the total solar energy striking earth today would have been available 4 billion years ago. Miller and Orgel have pointed out that although 0°C would give a better chance for the accumulation of sufficient concentrations of organic compounds in the ocean, −21°C would be ideal for chemical evolution to be most reasonable. At −21°C, however, (it is not unlikely) the ocean would be frozen. Such temperatures would give significantly longer half-lives to organic compounds. A solid reaction medium is much less favorable for synthesis than a liquid one, however, which could only have prevailed in equatorial regions.

Temperatures would have been some 20°C lower than today unless the "Greenhouse Effect" of the primitive atmosphere were much more efficient than the present one.[23] According to the Greenhouse Effect water vapor in the atmosphere transmits most of the solar energy to the earth's surface, which then re-emits energy at a longer wavelength in the infrared region of the spectrum. Instead of radiating off the planet, however, the re-emitted energy is absorbed by the water vapor, thus causing an elevated temperature. A lower temperature at the earth's surface would mean less water vapor in the atmosphere, hence a reduced Greenhouse Effect. Unless greater quantities of some other infrared absorbing material such as methane and especially ammonia were present in the early atmosphere, surely the average temperature of the earth would have been even more than 20°C lower than now, perhaps allowing a completely frozen ocean.[24] This prospect would seem probable because of the objection raised earlier against a substantial methane-ammonia primitive atmosphere.

The idea of a frozen ocean, which stems from astronomy, is not compatible with the view from geology that the earth was too hot 3.98 billion years ago and earlier to support life. Neither of these views can be held without some mechanism to account for a geologically rapid (less than 200 million years) decrease in temperature. This figure of less than 200 million years is based on the date of 3.81×10^9 years for the first fossil evidence of life, as cited by Brooks and Shaw.[25]

Lightning. It has usually been assumed that electrical activity on the primitive earth would have been comparable to that of today. If the early earth were some 20°C cooler than today because of less solar luminosity, however, it would significantly reduce thunderstorms on the earth, perhaps by a factor of 100 or more.[26] Atmospheric electrical storms arise under conditions which require minimally that water be evaporated and transported upward, an energy-consuming process. For thunderstorms to occur the air must be warm and humid below, and cold and dry above. It follows that at

20°C or more below present surface temperatures thunderstorm activity will be less, which is illustrated by the fact that not many thunderstorms occur in the Arctic, where less thermal energy is available to evaporate the water. With fewer electrical storms, lightning would be a far less abundant energy source than is generally believed, and it is generally believed anyway to have been a minor energy source. Sparks have been used as an energy source in laboratory experiments primarily as a matter of convenience.

Shock Waves. If there had been substantially fewer electrical storms due to a lower temperature on the early earth, it follows that thunder shock waves were less frequent as well. Shock waves would also result, however, from the impact of meteors passing through the atmosphere. Nevertheless, as table 3-1 shows, the meteorite contribution to the energy supply was less than a tenth of the energy supplied by electrical discharges. Total energy available from shock waves in any event was more than a thousand times less abundant than ultraviolet light. The optimism over shock waves as a candidate for a major energy source arises, not from its abundance, however, but from its efficiency. Shock waves are considered more than a million times more efficient than ultraviolet in producing amino acids.[27] Thus the "unexpected conclusion" is reached that shock waves may very well have been *the* principal energy source for prebiotic synthesis on the early earth by a factor of a thousand.[28] Such optimism regarding possible shock-wave synthesis should be tempered by what we shall call the "Concerto Effect". This term means that all the energy sources (and chemicals) act together or in concert in the natural situation—both in synthesis and in destruction of organic compounds. One energy source destroys what another source produces. Since these sources are quite generally more effective in destruction than in synthesis, this amounts to a preponderance of destruction. Amino acids produced in the atmosphere by electrical discharges or shock waves, for example, would be vulnerable to long-wavelength (> 2000 Å) ultraviolet photodissociation, which we mentioned earlier. This is a major objection to the accumulation of amino acids in the primitive ocean. The problem posed by the Concerto Effect will remain even if the dispute concerning the temperature history of the earth is resolved. Synthesized organic molecules are quite defenseless and vulnerable to destruction by all the energy sources.

Hydrolysis of HCN and Nitriles (RCN)

According to Ponnamperuma, hydrogen cyanide may be "the most important intermediate leading to the origin of life."[29] It is an ingredient for the production of amino acids in the Strecker synthesis (see Chapter 3). It also is considered a starting material in the synthesis of adenine and a host of other biomolecules, as shown in figure 4-1. The value of HCN in the chemical evolution scenario is enhanced by the fact that it escapes rapid destruction in the atmosphere by ultraviolet irradiation.[30] Hydrogen cyanide would have been generated in the atmosphere primarily by electrical discharges and collected in the ocean. It is the ubiquitous water molecule, however, that is the main obstacle to the reaction involving HCN and its nitrile derivatives.[31] For example, HCN adds water to its triple bond to form formamide, which, upon further hydrolysis, produces formic acid.

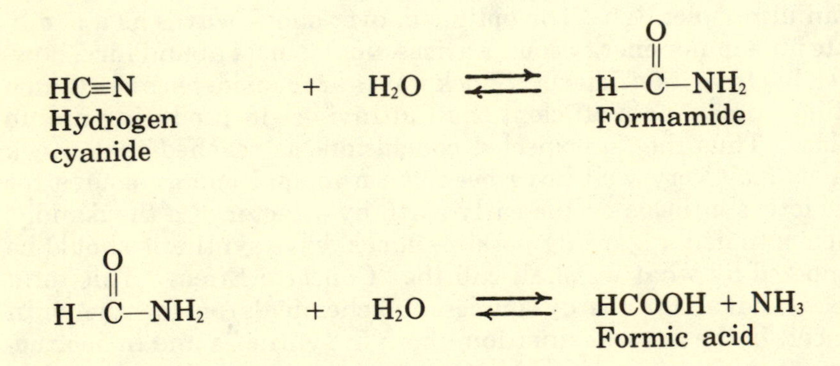

Formic acid is the major product in electrical discharge experiments, and this reaction probably accounts for that fact. As long as HCN concentrations are 0.01M or less, hydrolysis predominates. As we shall discuss later, HCN polymerization will predominate in more concentrated solutions. But there are problems. "Such a high steady-state concentration in an extended water mass does not seem likely since the hydrolysis to formic acid requires at most a very few years at reasonable pH's and temperature."[32] The highest average concentration of HCN would have been 10^{-6}M.[33] In other words, it is very

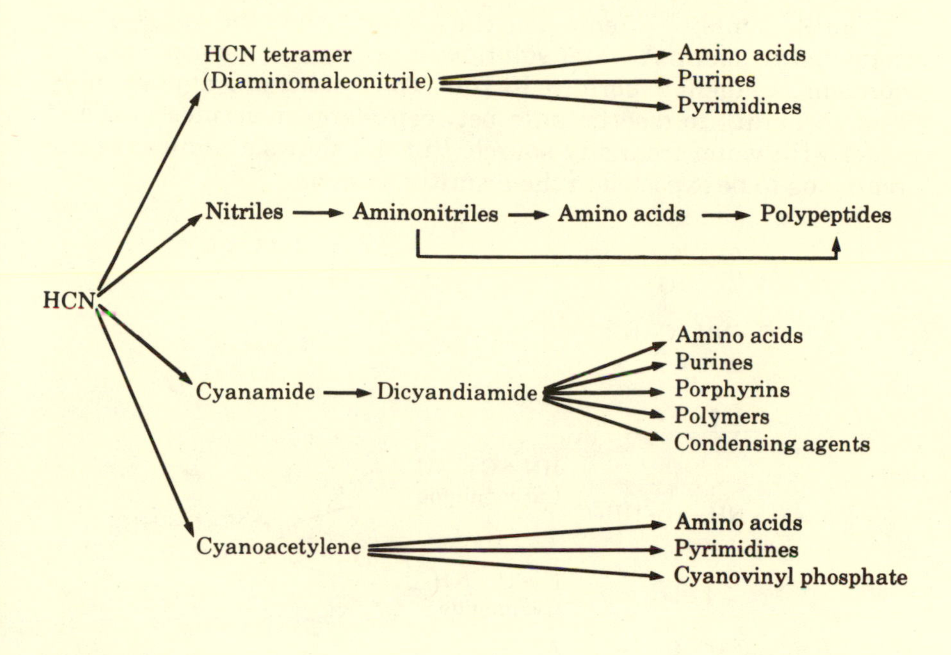

Figure 4-1.
Chemical evolution of biomolecules from HCN.

unlikely that HCN could have played a significant role in the synthesis of biologically meaningful molecules in an oceanic chemical soup. This is significant since many recent scenarios give HCN a prominent place. Also, a variety of HCN-derived nitriles have been suggested as having an important role as condensing agents in the synthesis of biologically significant polymers.[34] Examples of condensing agents include cyanogen, cyanamide, dicyanamide, and cyanoacetylene. Some of these were mentioned in the review of ocean experiments in Chapter 3. The ease with which these cyano compounds enter into reaction with water is, however, a major barrier to their usefulness in synthesis. It is, of course, the ability of these cyano compounds to react with water that makes them attractive candidates as condensing agents. The role of a condensing agent is to remove the water that is spilt-out or produced as a by-product in polymer formation. For example, when two amino acids react to form a dipeptide, a water molecule is released. Although dimer formation is thermodynamically unfavorable it can be made favorable simply by removing the water; hence the value of the condensing agent. A water solution, however, is a poor place for a condensing agent to perform its role. The condensing agent simply has no ability to discriminate between water molecules and will react with water from any source. Fig. 4-2 shows a number of the reactions to be expected in the primitive ocean.

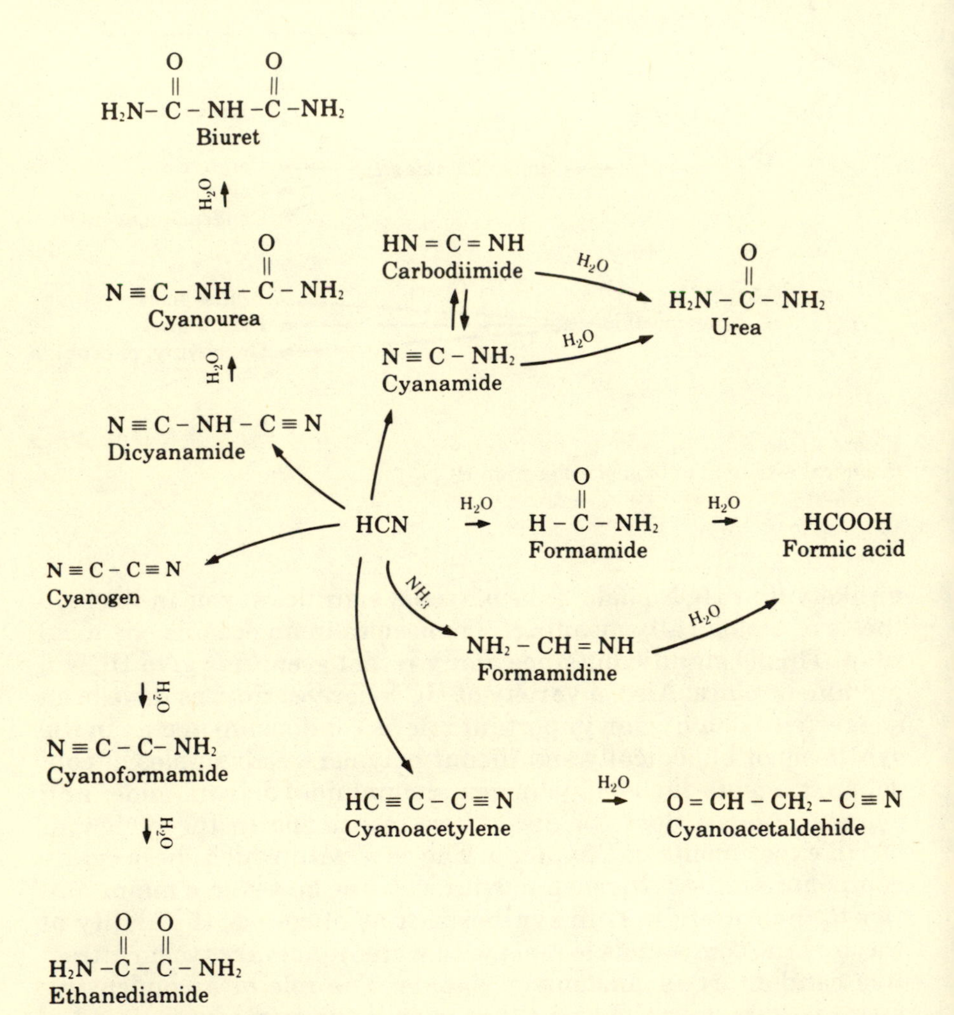

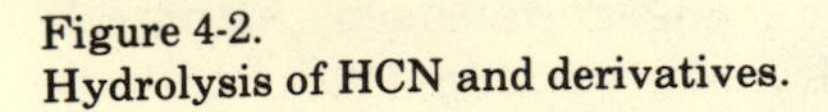

Figure 4-2.
Hydrolysis of HCN and derivatives.

Reaction of Carbonyl Group with Amino Group

The reaction of compounds containing a free amino group ($-NH_2$) with compounds containing a carbonyl group ($>C{=}O$) would have been a very important destructive process. This reaction would vastly diminish concentrations of important organic compounds in the primitive ocean. It can be written generally as follows:

$$\underset{\substack{\text{Carbonyl}\\\text{group}}}{>C{=}O} \quad + \underset{\substack{\text{Amino}\\\text{group}}}{H_2N-} \longrightarrow \left[\begin{array}{c} | \\ -C-NH- \\ | \\ OH \end{array}\right] \longrightarrow \underset{\text{Imine}}{>C{=}N- + H_2O}$$

Since the addition product (in brackets) is often unstable and loses water, this reaction is frequently called a *dehydration-condensation* reaction.

Many substances used in prebiotic simulation experiments (see Chapter 3) presumably would have been present in the oceanic soup. According to the general equation above, the amino group ($-NH_2$) of amines (including the free amino group in purines and pyrimidines) and amino acids would combine with the carbonyl group ($>C{=}O$) of reducing sugars, aldehydes, and a few ketones. Huge amounts of essential organic compounds would thus be removed from the soup by these reactions.[35]

These reactions would have greatly diminished not only amino acid concentration but also the concentration of aldehydes. Buildup of concentrations of aldehydes, especially formaldehyde, would have been important in the primordial synthesis of sugars. Polymerization of formaldehyde in alkaline solution has given a variety of sugars vital to life, including glucose, ribose, and deoxyribose. Studies of thermodynamic and kinetic stability of the important sugars suggest, however, that only insignificant amounts of them could have existed in the primordial ocean.[36] Add to this the chemical reality of reactions of sugars with amino compounds and the problem is seen as acute. Such low sugar concentrations argue strongly against formation of nucleic acids since they contain sugar.

Indiscriminate Amide Synthesis in Making Polypeptides

In the amide synthesis reaction the amino group ($-NH_2$) of amino acids would displace the hydroxyl group ($-OH$) from carboxylic acids (RCOOH) including amino acids. This is the reaction which occurs between amino acids to produce polypeptides and proteins. For example, two amino acids may combine to form a dipeptide:

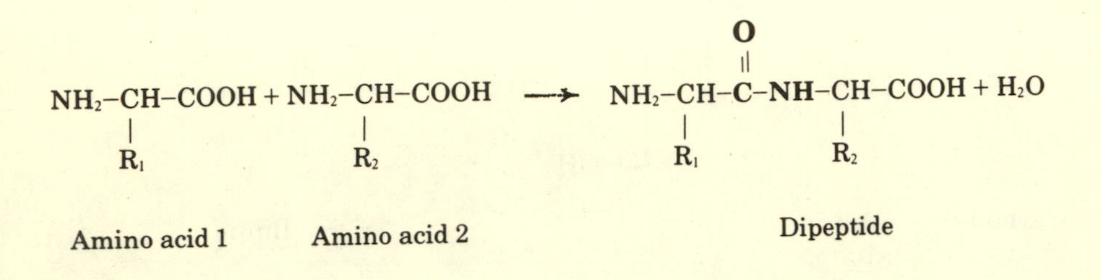

Because two molecules are combined with the release of water this is also called a *dehydration-condensation* reaction. According to most chemical evolution scenarios this reaction probably accounted for the primordial synthesis of polypeptides and proteins. There would, however, have been many different kinds of amino acids in the soup available for reaction. Most of these would have been non-proteinous. For example, results from Miller's spark discharge experiments (table 4-1) show many more non-proteinous than proteinous amino acids. In most cases more than one isomer (molecules with the same number of atoms but different geometry) is found for a given empirical formula. For example, three amino acid isomers are formed with formula $C_4H_9NO_3$, two of which are non-proteinous. All eight isomers of formula $C_4H_9NO_2$ are non-proteinous (fig. 4-3). It is obvious that something other than availability determines the selection of the set of 20 amino acids used in contemporary proteins. In addition, the amino acids produced in these experiments form a racemic mixture—an equal amount of both D- and L-amino acids. Proteinous and non-proteinous amino acids, both D- and L-, would lead to an indiscriminate production of polypeptides. These polypeptides would have scarce resemblance to protein. Protein not only requires exclusive use of L-amino acids, but also the use of a particular subset of only 20 amino acids. In addition, a biofunctional protein requires a

Table 4-1.
Yields of amino acids obtained from sparking a mixture of CH_4, NH_3, H_2O and H_2.

Compound	Relative Yield	Empirical Formula
Alanine	1000	$C_3H_7NO_2$ (1/3)
Glycine	557	$C_2H_5NO_2$ (1/1)
α-Amino-n-butyric acid	342	$C_4H_9NO_2$ (1/8)
α-Hydroxy-γ-aminobutyric acid	94	$C_4H_9NO_3$ (1/3)
Norvaline	77	$C_5H_{11}NO_2$ (1/7)
Sarcosine	70	$C_3H_7NO_2$ (2/3)
Aspartic acid	43	$C_4H_7NO_4$ (1/1)
α, γ-Diaminobutyric acid	42	$C_4H_{10}N_2O_2$ (1/1)
N-Ethylglycine	38	$C_4H_9NO_2$ (2/8)
α-Aminoisobutyric acid	38	$C_4H_9NO_2$ (3/8)
Valine	25	$C_5H_{11}NO_2$ (2/7)
β-Alanine	24	$C_3H_7NO_2$ (3/3)
N-Methylalanine	19	$C_4H_9NO_2$ (4/8)
Leucine	14	$C_6H_{13}NO_2$ (1/5)
Glutamic acid	10	$C_5H_9NO_4$ (1/1)
α, β-Diaminopropionic acid	8	$C_3H_8N_2O_2$ (1/1)
Norleucine	8	$C_6H_{13}NO_2$ (2/5)
Isoserine	7	$C_3H_7NO_3$ (1/2)
Alloisoleucine	6	$C_6H_{13}NO_2$ (3/5)
Serine	6	$C_3H_7NO_3$ (2/2)
Isovaline	6	$C_5H_{11}NO_2$ (3/7)
N-Methyl-β-alanine	6	$C_4H_9NO_2$ (5/8)
Isoleucine	6	$C_6H_{13}NO_2$ (4/5)
γ-Aminobutyric acid	3	$C_4H_9NO_2$ (6/8)
N-Propylglycine	~3	$C_5H_{11}NO_2$ (4/7)
N-Isopropyglycine	~3	$C_5H_{11}NO_2$ (5/7)
N-Ethyl-β-alanine	~3	$C_5H_{11}NO_2$ (6/7)
Proline	2	$C_5H_9NO_2$ (1/1)
Threonine	~1	$C_4H_9NO_3$ (2/3)
Allothreonine	~1	$C_4H_9NO_3$ (3/3)
β-Amino-n-butyric acid	~0.4	$C_4H_9NO_2$ (7/8)
β-Amino-isobutyric acid	~0.4	$C_4H_9NO_2$ (8/8)
N-Ethylalanine	<0.3	$C_5H_{11}NO_2$ (7/7)
Pipecolic acid	~0.06	$C_6H_{11}NO_2$ (1/1)
tert-Leucine	<0.03	$C_6H_{13}NO_2$ (5/5)

(After S. Miller, 1974. *Origins of Life* **5**, 139)
Yields are relative to alanine and presented in descending order. Numbers in parentheses indicate the comparative abundance of each compound among its isomers. For example, alanine is the most abundant of three isomers with the empirical formula $C_3H_7NO_2$. Biologically relevant amino acids are written in italics.

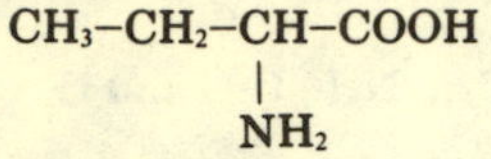

$CH_3-CH_2-CH-COOH$
|
NH_2

α-Amino-n-butyric acid

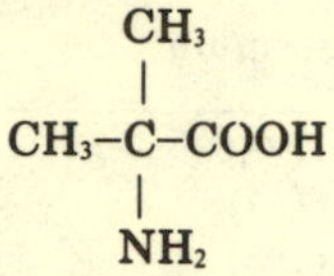

CH_3
|
$CH_3-C-COOH$
|
NH_2

α-Aminoisobutyric acid

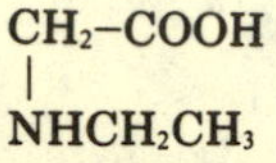

CH_2-COOH
|
$NHCH_2CH_3$

N-Ethylglycine

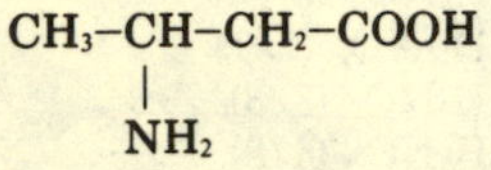

$CH_3-CH-CH_2-COOH$
|
NH_2

β-Amino-n-butyric acid

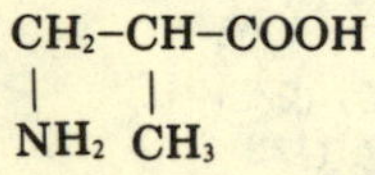

$CH_2-CH-COOH$
| |
NH_2 CH_3

β-Aminoisobutyric acid

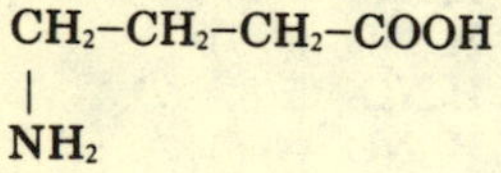

$CH_2-CH_2-CH_2-COOH$
|
NH_2

γ-Aminobutyric acid

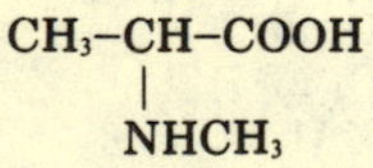

$CH_3-CH-COOH$
|
$NHCH_3$

N-Methylalanine

CH_2-CH_2-COOH
|
$NHCH_3$

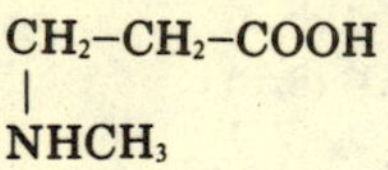

N-Methyl-β-alanine

Figure 4-3.
Structural isomers of amino acids with empirical formula $C_4H_9NO_2$ found in Miller experiment. None are found in proteins.

precise sequence of the amino acids. The important fact that amino acids do not combine spontaneously, but require an input of energy, is a special problem discussed in detail in Chapters 8 and 9.

Termination of Polypeptides and Polynucleotides

If the various dilution processes considered so far had not prevented formation of polypeptides and polynucleotides, these macromolecules would certainly have been vulnerable to degradation by chemical interaction with a variety of substances in the ocean. We have already seen how amino acids in the oceanic chemical soup would be expected to react with a variety of chemicals. In a similar fashion, growing polypeptides would be terminated by reactions with amines, aldehydes, ketones, reducing sugars* or carboxylic acids. If by some remote chance a true protein did develop in the ocean, its viability would be predictably of short duration. For example, formaldehyde would readily react with free amino groups to form methylene cross-linkages between proteins.[37] This would tie up certain reactive sites, and retard the reaction of protein with other chemical agents. To illustrate, "irreversible combination of formaldehyde with asparagine amide groups" would result in a compound which is "stable to dilute boiling phosphoric acid."[38] This tying up process is the principle of the well-known tanning reaction, and is used similarly to retard cadaver decay. "In general, reaction with formaldehyde hardens proteins, decreases their water-sensitivity, and increases their resistance to the action of chemical reagents and enzymes."[39] Survival of proteins in the soup would have been difficult indeed.

If we assume some small amount of nucleic acids formed in the primitive ocean, they too would be vulnerable to immediate attack by formaldehyde, particularly at the free amino groups of adenine, guanine, and cytosine. Some of the bonds formed with nucleic acids would be so stable that hydrolysis to liberate free formaldehyde would take place only by boiling with concentrated sulfuric acid.[40] As with proteins, it is difficult to conceive of a viable nucleic acid

---

*It is interesting to note that in certain abnormal situations, such as diabetes, the carbonyl group of glucose will form chemical bonds with the amino group of cellular proteins, a process called glycosylation. (See A.L. Notkins, 1975. *Sci. Amer.* **241**, 62.)

existing in the primordial soup for more than a very brief period of time.

Hydrolysis of Amino Acids and Polypeptides

But what if polypeptides and other biopolymers had formed in the prebiotic soup? What would their fate have been? In general the half-lives of these polymers in contact with water are on the order of days and months—time spans which are surely geologically insignificant.[41]

Besides breaking up polypeptides, hydrolysis would have destroyed many amino acids.[42] In acid solution hydrolysis would consume most of the tryptophan, and some of the serine and threonine. Further, acid hydrolysis would convert cysteine to cystine, and would deamidate glutamine and asparagine. On the other hand, hydrolysis would destroy serine, threonine, cystine, cysteine, and arginine in the alkaline solution generally regarded to have characterized the early ocean. An alkaline solution would also have caused several deamidations.

Precipitation of Fatty Acids and Phosphate with Calcium and Magnesium Salts

We have already discussed how attenuated concentrations of the nucleic acids in the primitive ocean would have been. Another reason for this is the severe restriction caused by the poor solubility of phosphate, an essential ingredient of nucleic acids. No soluble phosphates are known that could plausibly have existed in the primitive ocean.[43] They would be expected to precipitate out of the soup by forming insoluble salts with calcium and magnesium ions.[44] For example, hydroxylapatite, $Ca_5(PO_4)_3OH$, has a solubility product of about 10^{-57}. Since there would have been ample amounts of dissolved calcium in the soup it is difficult to imagine a phosphate concentration greater than 3×10^{-6}M.[45] As Griffith et al., have noted, "the primitive seas were probably severely deficient in phosphorus."[46] In addition, fatty acids which are essential ingredients for synthesis of cell membranes would have precipitated out of the soup by forming insoluble salts with magnesium and calcium ions.[47]

Adsorption of Hydrocarbons and Organic Nitrogen Containing Compounds on Sinking Clay Particles

If there is any merit to the view that methane was an important constituent of the primitive atmosphere, hydrocarbons surely must have formed in the atmosphere under the influence of ultraviolet irradiation and fallen into the ocean.[48] Hydrocarbons would then be brought to rest on the ocean bottom by adsorption on sedimenting clays. The earliest Precambrian deposits would be expected to contain unusually large proportions of hydrocarbon material or its carbon remains. They do not, however.[49]

Nitrogen-containing organic compounds would also be expected to have been removed from the ocean by adsorption on clay particles. As Nissenbaum has noted, "We have also no reason to doubt that ...adsorption on mineral surfaces, and especially clays, was operative in those remote times."[50] Brooks and Shaw have said in *Origin and Development of Living Systems*:

> If there ever was a primitive soup, then we would expect to find at least somewhere on this planet either massive sediments containing enormous amounts of the various nitrogenous organic compounds, amino acids, purines, pyrimidines, and the like, or alternatively in much-metamorphosed sediments we should find vast amounts of nitrogenous cokes [graphite-like nitrogen-containing materials]. In fact *no such materials have been found anywhere on earth*.[51] (Emphasis added.)

In summary, the above dilution processes operating in both the atmosphere and in the ocean would have greatly diminished concentrations of essential precursor chemicals. Although these processes have been identified and discussed individually, they would have worked synergistically, or in concert. Fig. 4-4 summarizes the Concerto Effect by using many of the individual reactions discussed in this chapter. It seems probable that in an oceanic chemical soup the synthesis of RNA and other essential biomolecules would have been short-circuited at nearly every turn by many cross-reactions. The overall result would be very small steady-state concentrations of important soup ingredients.

Concentration of Essential Chemicals in the Prebiotic Soup: The Example of Amino Acids

The picture emerges of a primitive earth with oceans much more dilute in organic material than is often assumed. How dilute would the early oceans have been? We shall now develop a more quantitative estimate of the concentration of important ingredients in the primitive oceanic soup.

It is a widely held view that the early oceans would have contained huge quantities of organic material. Urey theorized the primitive ocean was rich in organic compounds, containing enough dissolved carbon compounds to make perhaps a 10% solution.[52] This is equivalent to a concentration of 10^{-3}M for each of 1000 chemical compounds in a soup, with an average molecular weight of 100 for each compound.

More recent estimates have revised Urey's estimate downward. Sagan suggests that a 0.3% to 3% solution would result from dissolving in the oceans the organic matter produced by ultraviolet irradiation of a primitive atmosphere for a billion years.[53] Based on data from electric discharge experiments, Wolman et al., have estimated that the oceans of the primitive earth would have been about 2×10^{-3}M in amino acids.[54] Both of these revised estimates are extremely optimistic however. Sagan's estimate acknowledges "no destruction of synthesized material,"[55] and Wolman et al. "assume...that decomposition of amino acids after synthesis was minimal."[56] To the contrary, as much of this chapter has shown, any realistic assessment of the fate of chemicals such as amino acids on the early earth *cannot ignore* their very considerable destruction either by energy sources or by chemical interaction in the soup.

The effectiveness of these various natural processes to destroy organic products suggests that the steady-state concentration of amino acids in the primitive oceans would have been quite low. Just how low can only be estimated in ways involving much uncertainty. Nevertheless, plausible estimates which take into account the destructive processes have been made. One estimate by Dose considers ultraviolet destructive effects in the ocean, but ignores both ultraviolet destruction of amino acids in the atmosphere and the destructive interaction between amino acids and other chemicals in the

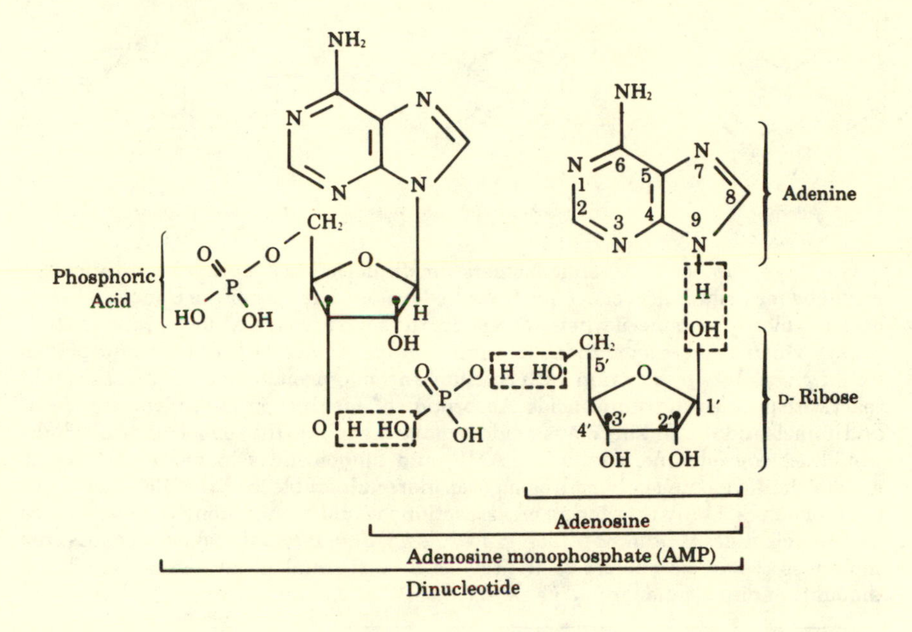

Figure 4-4.
The role of the Concerto Effect in the formation of dinucleotide in the prebiotic soup.

Assume initially an aqueous soup consisting of adenine, D-ribose and phosphoric acid. There are 3 sites on adenine (N7, N9 and NH_2 attached to C6) which can react with hydroxyl at 5 sites on D-ribose (C1′α, C1′β, C2′, C3′, and C5′) which gives rise to 15 structural isomers of adenosine. Only one of these, i.e., 9 (1 -β-D-ribofuranosyl) adenine, is found in living things. Proceeding to the level of AMP (adenosine monophosphate) there are 3 possible sites of attachment of phosphate to D-ribose (C2′, C3′, and C5′). Consequently the number of structural isomers of AMP (adenosine monophosphate) are the number for adenosine times 3, or 45. At the dinucleotide level, since there are still 2 free -OH groups on D-ribose, the number of possible isomers would be that of AMP times 2, or 90.

Although C2′ and C3′ of ribose are chiral carbons, and the hydroxyls attached to them may be conceived to be in four different arrangements, note that by definition only one of these is called ribose. The other sugar arrangements are given different names, i.e., lyxose, xylose, and arabinose. In general then, the pentose sugars have 8 isomers (D- and L-). Consequently, the total number of dinucleotide isomers would be determined as:

3 (sites on adenine) x 5 (sites on D-ribose)
x 8 (pentoses) x 3 (sites left on pentose for phosphate links)
x 2 (sites left on pentose for dinucleotide links) = 720.

Also observe that aminopurines can form with the -NH_2 at C2 or C8 as well as at the C6 position for adenine: The number of possible isomers of dinucleotide would now be as previously determined times 3, or 2160.

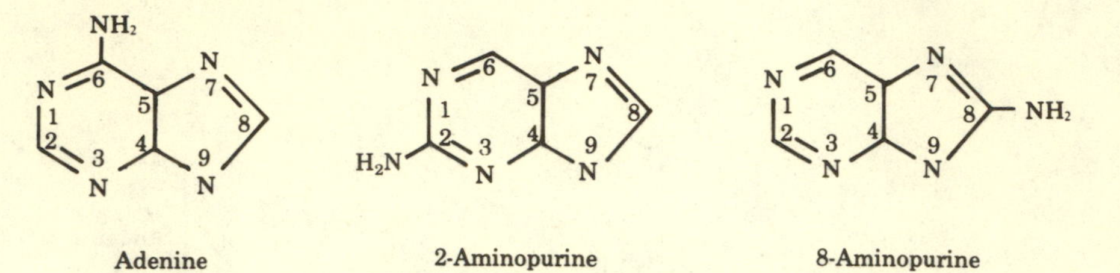

Adenine 2-Aminopurine 8-Aminopurine

The large number of possible isomers for dinucleotide suggests how difficult it would be for meaningful concentrations to develop. The role of the Concerto Effect becomes more pronounced when we consider the soup to contain aldehydes and other sugars which could react with the free amino group of adenine (purines). Phosphates would precipitate by reaction with calcium and magnesium salts. Pentoses would react with amines and amino acids. Absorption of adenine (purine), adenosine, AMP and dinucleotide on sinking clays would remove them from the soup. Ultraviolet light would destroy adenine, adenosine, AMP, and dinucleotides in the upper surface waters; the developing polymers being even more vulnerable to ultraviolet decay than the monomers. Many interfering crossreactions would occur among the nucleotides and dinucleotides to terminate their growth. And of course all the substituent organic molecules would be subject to hydrolysis and thermal decay. Extremely small amounts of dinucleotide would be expected.

---

ocean.[57] Dose arrives at an *upper limit* estimate of amino acid concentration in the primitive ocean of 10^{-7}M, some 10,000 times more dilute than the optimistic estimates reported above. As it turns out, the present-day average concentration of amino acids in the North Atlantic Ocean is also about 10^{-7}M.[58]

A second estimate which gives a similar result considers the destructive interaction between amino acids and various soup ingredients, especially sugars, but ignores the ultraviolet destruction process entirely.[59] This estimate is based on a process of scavenging amino acids from the soup followed by polymerization. After a complicated polymerization reaction, the polymer is removed by sedimentation. The first step of the polymerization process involves a dehydration-condensation reaction between the amino group ($-NH_2$) of amino acids and the carbonyl group ($>C=O$) of reducing sugars, as previously discussed. In this manner the oceans of today are scavenged of their sugars and amino acids which come indirectly from the decay of more complex organic matter of previously living things. The early ocean, on the other hand, would have been directly supplied with abiotically derived amino acids and sugars. There is no reason to doubt the operation of the scavenging process in the early oceans.

Since this process if geologically instantaneous (1000-3500 years) it is difficult to imagine the primitive soup ever more concentrated than 10^{-7}M with amino acids.[60] Nissenbaum et al., have summed up the importance of the scavenging process by observing:

> This scavenging of dissolved organic matter from the oceans by polymerization and sedimentation would have left the oceans much more depleted in abiotically formed organic material than is usually assumed. It is difficult to see how, under such conditions, the "primordial soup" could have existed at all.[61]

A third estimate of amino acid concentration in the early oceans considers ultraviolet destruction both in the atmosphere *and* in the oceans, but ignores the destructive reactions of amino acids with other soup ingredients.[62] This estimate is based on a comparison of rates of formation of amino acids versus their decomposition by ultraviolet. It shows that only about 3% of the amino acids produced in the upper atmosphere (where most UV-promoted amino acid synthesis would have occured) could have safely passed to the ocean. This would yield a maximum steady-state amino acid concentration of 10^{-12}M in the primitive ocean.

A truly realistic estimate must combine these factors and other destructive processes, and consider the effects of all the energy sources as well. It would be a very difficult estimate to make. Even so, the above estimates are sufficient to suggest that the primordial ocean would have been an extremely dilute "soup," much too dilute to reasonably expect the spontaneous formation of proteins.[63] Although the notion persists at the popular level that life began in the ocean, among scholars and researchers in the field, "it is now generally accepted that the concentration of the soup was probably too small for efficient synthesis, particularly of biopolymers."[64]

We conclude that if there ever was a prebiotic oceanic soup of chemicals, it would have been too dilute for chemical evolution rates to have been significant.

Concentrating Little Ponds

The realization that an organic soup would have been too dilute for direct formation of polymers may seem devastating to chemical evolution views. However, as Bernal has written, "The original concept of the primitive soup must be rejected only in so far as it applies to oceans or large volumes of water, and interest must be transferred to reactions *in more limited zones*."[65] (Emphasis added). By this he meant lakes, pools, lagoons, and the like. These more limited zones might then have been the locus of life's origin rather than the ocean. The significance of these local places is their associated mechanisms for concentrating essential chemicals. By concentrating the monomers, the probability of their molecular interaction would have been increased, thus increasing reaction rates according to the law of mass action. This law states that the rate of a chemical reaction is directly proportional to the concentration of the reacting substances. Hence in concentrated ponds the probability of polymer formation would have been considerably enhanced.

Even phosphate, which was previously mentioned as limited to a concentration of about 10^{-6}M in the ocean, might conceivably be concentrated in a pool deficient in calcium and magnesium salts. A means to increased phosphate concentration seems essential, since the phosphorylating process to activate amino acids for further reaction assumes those conditions. The suggestion is made plausible since natural deposits of $NaBePO_4$, a highly soluble phosphate, and even deposits of monosodium phosphate, NaH_2PO_4, have been found, probably arising from non-biological processes.[66]

Two mechanisms for concentrating organic chemicals in lakes, pools, lagoons, etc. have been suggested. These are (1) simple evaporation and (2) freezing the body of water. Both of these concentrating mechanisms have been suggested as playing a significant role in enhancing chemical evolution rates.

Evaporation[67]

As a hypothetical evaporation mechanism (see fig. 4-5), let us picture a small pool in a cave (so the accumulating organic compounds are protected from ultraviolet light) located near a fumarole (so there is a heat source for evaporating the water) and so situated at the coast that at high tide the ocean soup will overflow into the pool to supply organic compounds without washing away the concentrated organics in the same action. Between high tides evaporation slightly increases the concentration of organic compounds. After many iterations of this cyclic process a reservoir of concentrated organic compounds is developed.

Although this hypothetical evaporation scheme is only one of many that can be envisioned, we shall use it to illustrate several facets of the mechanism. Whatever the details of the specific evaporating pool, lake, or lagoon, it must include:

1. A suitable reservoir for concentrating organic compounds.
2. A heat source for evaporating water.
3. The repeated admission of oceanic soup into the reservoir.
4. Some means to protect the organic compounds from ultraviolet light.

Figure 4-5.
A hypothetical evaporation mechanism.
A small pond in a cave protects accumulating organic compounds from ultraviolet light. Located nearby is a fumarole which evaporates the water between high tides. During high tides dilute organic soup refills the ponds, but without flooding away concentrated material.

If such evaporating pools existed they would surely have tended to concentrate non-volatile substances such as amino acids, purines, etc. But evaporating pools would have been inadequate for concentrating volatile substances such as aldehydes and HCN. Instead of concentrating volatile substances upon evaporation of the pool, they would simply evaporate and redissolve in more dilute water bodies. This is particularly important since, as we noted earlier, HCN will significantly polymerize only if it can be concentrated to more than 0.01M. Since HCN in the open ocean would have been on the order of 10^{-6}M,[68] it is clear that some other concentrating mechanism must have been involved if HCN were significant in chemical evolution.

Freezing

If the solar luminosity on the early earth was less than today, as previously discussed, then many of the water bodies of earth would have been covered with ice, if not completely frozen. In certain equatorial regions (where liquid water could have persisted) the water bodies might have alternately frozen and thawed with the seasons. In this setting Orgel has shown that dilute solutions of HCN at 10^{-5}M from the ocean might run into a localized pool in summer and collect there. As the water freezes over in winter, the HCN concentrates in the solution beneath the ice. A 10% conversion to organic material might occur. As this cyclic process continued, material of molar concentration might accumulate every million years.[69]

Critique of Concentrating Mechanism

There is no known geological evidence for organic pools, concentrated by these or other mechanisms, ever existing on this planet.[70] In contrast, much evidence is available that inorganic pools existed in early times. Such inorganic pools can be seen today at Yellowstone National Park.

It is not too significant, however, that evidence for isolated reservoirs of organic compounds has not been located. They would undoubtedly have been fewer in number, since requirements for an organic pool would have been more stringent. If evidence is available for such organic pools it may take some time to locate.

More significant is the fact mentioned earlier that geological evidence for the oceanic soup has not been located. If there ever was a dilute ocean that fed organic compounds into these smaller pools, there should be abundant evidence for it in the lower Precambrian sediments. None has been located, however. Remember, if the soup were as massive as the theory suggests, organic remains should be literally all over the earth in deep sediments of great age. Scientists have looked but have not found organic compounds.

Still, if by some means concentrated pools did develop, not only would the desired materials concentrate, but also the undesirable impurities. For example, an evaporating pond concentrating nonvolatiles such as amino acids would also concentrate sea salts such as NaCl.[71] A freezing pond concentrating volatile substances such as HCN would do the same. If such salts were in great excess (which is not unlikely), then organic compounds in the pond could not have been significantly concentrated as a result of the "salting-out effect." This effect assumes the NaCl and other sea salts compete for the water molecules in the solution of organic compounds such as amino acids. Salt has greater affinity for water than do these organic compounds. Therefore, in order for the salt to be dissolved the organic compounds must precipitate out of solution.

It is another type of "impurity," however, that would have been the greatest obstacle to the successful concentration of organic compounds in limited zones. This would be the host of oceanic organic compounds such as amines, amino acids, aldehydes, ketones, sugars, carboxylic acids, etc. that would have destructively interacted in the ocean.[72] The usual consequences of concentrating these would be, according to the law of mass action, merely an acceleration of the many destructive reactions (as well as the constructive reactions) that would also occur at slower rates in the more dilute ocean, as already discussed.

Hydrogen cyanide would seem to be an exception, since on concentration, polymerization tends to predominate. Hydrolysis of HCN would predominate in the dilute ocean. Polymers of HCN, however, would yield the vulnerable amino acids upon hydrolysis.[73] If peptides formed directly from HCN polymerized in the atmosphere and fell into the ocean,[74] these would be terminated by reacting with amines, carboxylic acids, etc., as discussed earlier.

Concentrating mechanisms have occupied the attention of some investigators. Stemming from this discussion, however, it is our observation that what is needed is a natural *sorting* mechanism. The problem demands a means of selecting organic compounds and isolating them from other chemicals with which they could destructively interact. Yet there is nothing (but the need) to suggest that such a sorting mechanism ever existed on this planet.

In other words, for these more limited zones (e.g., lakes, pools, lagoons), as for the ocean itself, it is difficult to imagine significant concentrations of essential organic compounds ever accumulating. As we have seen, degradative forces need to be taken into account in realistic estimates of concentrations, and they have frequently been ignored.

Conclusion

Based on the foregoing geochemical assessment, we conclude that both in the atmosphere and in the various water basins of the primitive earth, many destructive interactions would have so vastly diminished, if not altogether consumed, essential precursor chemicals, that chemical evolution rates would have been negligible. The soup would have been too dilute for direct polymerization to occur. Even local ponds for concentrating soup ingredients would have met with the same problem.

Furthermore, no geological evidence indicates an organic soup, even a small organic pond, ever existed on this planet. It is becoming clear that however life began on earth, the usually conceived notion that life emerged from an oceanic soup of organic chemicals is a most implausible hypothesis. We may therefore with fairness call this scenario "the myth of the prebiotic soup."

References

1. P.H. Abelson, 1966. *Proc. Nat. Acad. Sci. U.S.* **55**, 1365; P.E. Cloud, 1968. *Science* **160**, 729.
2. A.C. Lasaga, H.D. Holland, and M.J. Dwyer, 1971. *Science* **174**, 53.
3. J.P. Pinto, G.R. Gladstone, Y.L. Yung, 1980. *Science* **210**, 1983; C. Ellis and A.A. Wells, 1941. *The Chemical Action of Ultraviolet Rays*, revised and enlarged edition by F.F. Heyroth. New York: Reinhold, p. 417; Abelson, *Proc. Nat. Acad. Sci. U.S.* 1365; H.R. Hulett, 1969. *J. Theoret. Biol.* **24**, 56; H.R. Hulett, 1973. In *Proceedings of the Fourth Conference on Origins of Life: Chemistry and Radioastronomy*, ed. Lynn Margulis. New York: Springer-Verlag, p. 80.
4. N.H. Horowitz, F.D. Drake, S.L. Miller, L.E. Orgel, C. Sagan, "The Origins of

Life," 1970. In *Biology and the Future of Man*, ed. P. Handler. New York: Oxford U. Press, p. 163.
5. Abelson, 1966. *Proc. Nat. Acad. Sci. U.S.*, 1365; J.P. Ferris and D.E. Nicodem, 1972. *Nature* **238**, 268. J.P. Ferris and D.E. Nicodem, 1974. In *The Origin of Life and Evolutionary Biochemistry*, ed. K. Dose, S.W. Fox, G.A. Deborin, and T.E. Pavlovskaya. New York: Plenum Press, p. 107; W.R. Kuhn and S.K. Atreya, 1979. *Icarus* **37**, 207; J.S. Levine, 1982. *J. Mol. Evol.* **18**, 161.
6. Abelson, *Proc. Nat. Acad. Sci. U.S.*, 1365.
7. C. Sagan, 1977. *Nature* **269**, 224.
8. Kuhn and Atreya, *Icarus*, 207.
9. Ferris and Nicodem, *Nature*, 268; Ferris and Nicodem, in *The Origin of Life and Evolutionary Biochemistry*, p. 107; S. Miller, H. Urey, and J. Oro, 1976. *J. Mol. Evol.* **9**, 59.
10. Ferris and Nicodem, in *The Origin of Life and Evolutionary Biochemistry*, p. 107.
11. C.F. Davidson, 1965. *Proc. Nat. Acad. Sci. U.S.* **53**, 1194; R.T. Brinkman, 1969. *J. Geophys. Res.* **74**, 5355; E. Dimroth and M. Kimberly, 1976. *Can. J. Earth Sci.* **13**, 1161; J.C. Walton, 1976. *Origins* **3**, 66; J.H. Carver, 1981. *Nature* **292**, 136; J.S. Levine, 1982. *J. Mol. Evol.* **18**, 161.
12. News Release No. 30-72-7, Naval Research Laboratory, Washington, D.C.; G.R. Carruthers and Thornton Page, 1973. *Science* **177**, 788.
13. G.R. Carruthers, personal communication, Sept. 28, 1981.
14. Michael I. Ratner and James C.G. Walker, 1972. *J. Atmos. Sci.* **29**, 803; A.J. Blake and J.H. Carver, 1977. *J. Atmos. Sci.* **34**, 720; Carver, *Nature*, p. 136.
15. L.E. Orgel, 1973. *The Origins of Life*. New York: Wiley, p. 129; Horowitz, et al., *Biology*, p. 171.
16. Hulett, *J. Theoret. Biol.*, p. 60.
17. D.E. Hull, 1960. *Nature* **186**, 693.
18. Miller, Urey, and Oro, *J. Mol. Evol.*, p. 59.
19. Hull, *Nature*, 693.
20. L.V. Berkner and L.C. Marshall, 1965. *J. Atmos. Phys.* **22**, 225.
21. J.W.S. Pringle, 1954. *New Biology No. 16*, p. 54.
22. S.L. Miller and L.E. Orgel, 1974. *The Origins of Life on the Earth*. Englewood Cliffs, New Jersey: Prentice-Hall, p. 127.
23. Horowitz, et al., *Biology*, p. 174.
24. Horowitz, et al., *Biology*, p. 174.
25. J. Brooks and G. Shaw, 1973. *Origin and Development of Living Systems*. London and New York: Academic Press, p. 78.
26. Hulett in *Proceedings*, p. 93.
27. A. Bar-Nun, N. Bar-Nun, S.H. Bauer, and C. Sagan, 1970. *Science* **168**, 470.
28. Bar-Nun, et al., *Science*, p. 472.
29. C. Ponnamperuma, 1978. *Chemistry* **51**, 6.
30. Hulett, *J. Theoret. Biol.*, p. 61.
31. N.H. Horowitz and J.S. Hubbard, 1974. *Ann. Rev. Genetics* **8**, 393.
32. R. Sanchez, J. Ferris, and L.E. Orgel, 1966. *Science* **153**, 72.
33. Hulett, *J. Theoret. Biol.*, p. 61.
34. J. Hulshof and C. Ponnamperuma, 1976. *Origins of Life* **7**, 197.
35. A. Nissenbaum, 1976. *Origins of Life* **7**, 413.
36. Abelson, *Proc. Nat. Acad. Sci. U.S.*, p. 1365.
37. J.F. Walker, 1964. *Formaldehyde*, ACS Monograph 159, 3rd ed. New York: Reinhold, p. 399ff.

38. Ibid., p. 404.
39. Ibid., p. 399.
40. Ibid., p. 398.
41. K. Dose in *The Origin of Life and Evolutionary Biochemistry*, p. 69.
42. *Encyclopedia of Science and Technology*, 1982. "Amino Acids," vol. 1, p. 411-424. New York: McGraw-Hill.
43. S. Miller and M. Parris, 1964. *Nature* **204**, 1248; and D.H. Kenyon and G. Steinman, 1969. *Biochemical Predestination*. New York: McGraw-Hill, p. 175ff and 218.
44. Abelson, *Proc. Nat. Acad. Sci. U.S.*, p. 1365.
45. Hulett, *J. Theoret. Biol.*, p. 62.
46. E.J. Griffith, C. Ponnamperuma and N.W. Gabel, 1977. *Origins of Life* **8**, 71.
47. Abelson, *Proc. Nat. Acad. Sci. U.S.*, p. 1365.
48. Lasaga, Holland, and Dwyer, *Science*, p. 53.
49. Abelson, *Proc. Nat. Acad. Sci. U.S.*, p. 1365; Cloud, *Science*, p. 729.
50. Nissenbaum, *Origins of Life*, p. 415.
51. Brooks and Shaw, *Origin and Development of Living Systems*, p. 359.
52. H. Urey, 1952. *The Planets*. New Haven, Conn.: Yale Univ. Press, p. 152.
53. C. Sagan, 1961. *Rad Res.* **15**, 174.
54. T. Wolman, W.J. Haverland, and S. Miller, 1972. *Proc. Nat. Acad. Sci. U.S.* **69**, 809.
55. Sagan, *Rad. Res.*, p. 176.
56. Wolman, et al., *Proc. Nat. Acad. Sci. U.S.*, p. 811.
57. Dose in *The Origin of Life and Evolutionary Biochemistry*, p. 69.
58. R. Pocklington, 1976. *Nature* **230**, 374.
59. Nissenbaum, *Origins of Life*, p. 413.
60. Nissenbaum, *Origins of Life*, p. 413.
61. A. Nissenbaum, D.H. Kenyon, and J. Oro, 1975. *J. Mol. Evol.* **6**, 253.
62. Hull, *Nature*, p. 693.
63. Dose in *The Origin of Life and Evolutionary Biochemistry*, p. 74.
64. Nissenbaum, et al., *J. Mol. Evol.*, p. 259.
65. J.D. Bernal, 1960. *Nature* **186**, 694.
66. E.J. Griffith, C. Ponnamperuma, and N.W. Gabel, 1977. *Origins of Life* **8**, 71.
67. Miller and Orgel, *The Origins of Life on the Earth*, p. 129ff.
68. Hulett, *J. Theoret. Biol.*, p. 61.
69. R. Sanchez, J. Ferris, and L. Orgel, 1967. *J. Mol. Biol.* **30**, 223.
70. Dose in *The Origin of Life and Evolutionary Biochemistry*, p. 75.
71. Clair Edwin Folsome, 1979. *The Origin of Life*. San Francisco: W.H. Freeman and Co., p. 84; C.E. Folsome, ed. 1979. *Life: Origin and Evolution*. San Francisco: W.H. Freeman and Co., p. 3.
72. Folsome, *The Origin of Life*, p. 57, 59.
73. C.N. Matthews and R.E. Moser, 1967. *Nature* **215**, 1230.
74. Clifford N. Matthews, 1975. *Origins of Life* **6**, 155; C. Matthews, J. Nelson, P. Varma and R. Minard, 1977. *Science* **198**, 622.

CHAPTER 5

Reassessing the Early Earth and its Atmosphere

Over the past several decades, our growing understanding of the early earth has added crucial insight to theories of chemical evolution. In this chapter, three relevant points will be discussed. First, the time frame or the time available for chemical evolution will be established. Second, we will examine the chemical composition of the atmosphere on the primitive earth to determine if it was conducive to abiogenesis. Third, we will examine the important question of oxygen content on the early earth and in its atmosphere. This evaluation of plausible atmospheric conditions will help to establish constraints on the next generation of prebiotic simulation experiments. Many of the experiments reviewed in Chapter 3 assumed a strongly reducing primitive earth and atmosphere.

Establishing the Time Frame

One of the most dramatic changes in evolutionary theory since the 1960s has been in understanding the sharp reduction of the time available for abiogenic synthesis. As Richard E. Dickerson states,

"Perhaps the most striking aspect of the evolution of life on the earth is that it happened so fast."[1] In fact, Cyril Ponnamperuma of the University of Maryland and Carl Woese of the University of Illinois have suggested that life may be as old as the earth and that its origin may have virtually coincided with the birth of the planet.[2] In this section the data used to support such statements will be examined.

From radiometric dating techniques, the ages of stony meteorites have been set at 4.6 billion years.[3] If the sun, the planets, the meteorites, and other solar debris all formed from the same primordial dust cloud at about the same time, the earth would be approximately 4.6 billion years old. There exists a tremendous gap, however, in information about the earth from this date through the Precambrian until about 0.6 billion years ago.[4] This is especially so with respect to information about chemical evolution.[5] Until the late 1960s, the oldest suspected evidence for life was the occurrence of fossil stromatolites (photosynthesizing algae) in 2.7 billion-year-old limestone located in Southern Rhodesia.[6] However, in the late 1960s several scientists investigating very old rocks (3.2 billion years old) found evidence of molecular fossils and microfossils indicating past life.

Molecular Fossils

Molecular fossils (or chemical fossils) are actually chemical compounds found in the rocks and suspected of being the remains of once-living matter. The different types of chemicals that may indicate life are quite diverse. However, there are two different ways in which the compounds found may indicate an association with living organisms:

1. The compounds could be degradation products of chemicals found in living organisms. For example, isoprenoid alkanes (such as pristane and phytane) are assumed to result from the breakdown of chlorophyll. Isoprenoids found in ancient rocks could therefore be a record of living organisms. Many other chemicals associated with living organisms such as porphyrins and steranes, may be found in very old rocks as well.

2. During their metabolic processes, organisms selectively use carbon 12 over carbon 13. Thus, chemicals with a high carbon-12-to-carbon-13 ratio may indicate the occurrence of living processes.

Microfossils

Microfossils may also indicate past life. Microfossils are microscopic outlines in rocks indicating past life forms. Usually these are very simple algae-like spheroids or filaments found in carbon-rich rocks. It would be nice if some detail beyond their morphological characteristics were preserved for our inspection. This is rarely the case, however. Still, through the chemical analysis and microscopic examination of very old organic-rich rocks,* the whole field of chemical evolution has been changed dramatically. That is, before the identification of microfossils and molecular fossils, most scientists thought that perhaps as much as 2 billion years were available for chemical evolution to occur.

The Evidence

Since the 1960s, the following evidence has become available to support the view that life originated on the earth soon after its formation:

1. 1967: Micropaleontological studies of carbonaceous chert of the Fig Tree Series of South Africa (greater than 3.1 billion years old) indicated the presence of spheroidal microspheres. The photosynthetic nature of these primitive microorganisms was corroborated by organic geochemical and carbon isotopic studies.[7]

2. 1977: A population of organic walled microstructures from the Swaziland System, South Africa was identified as the morphological remains of primitive prokaryotes. The rocks were dated at 3.4 billion years old.[8]

3. 1979: Cell-like inclusions detected in the cherty layers of a quartzite, which is part of the Isua series in Southwest Greenland, consisted of biological materials. High carbon-12-to-carbon-13 ratios were found in the hydrocarbons. The age of the sequence is approximately 3.8 billion years.[9]

---

*This is tricky business, however, as sometimes inorganic materials can be mistaken for microfossils (E.L. Merek, 1973. *BioScience* **23**, 153; N. Henbest. 1981. *New Scientist* **92**, 164).

4. 1980: Researchers found biological-like cells in rocks from the "North Pole" region of Australia. The rocks were dated at 3.5 billion years old. Even more amazing was the fact that five different types of cells could be identified. "This tells us that life was diverse, abundant, and judging from the chemistry, really quite advanced."[10]

5. 1980: A fossilized mat of filamentous microorganisms called stromatolites have been preserved in ferruginous dolomitic chert of the Pilbara Block of Western Australia. They are estimated to be 3.4 to 3.5 billion years old.[11]

Until recently, "yeast-like microfossils" from the Isua belt in Southwest Greenland were regarded as evidence of living structures. Now, however, some researchers have raised questions about this interpretation,[12] suggesting that they are not the remains of early Archaen life forms. Thus the Australian deposits dating back to 3.5 billion years are currently considered the oldest sediments containing convincing evidence for biological activity. Even so, many scientists believe that life existed over 3.8 billion years ago.

The Time Available for Evolution

Brooks and Shaw state that the oldest rocks on earth are probably about 3.98 billion years old.[13] However, the oldest age confirmed by dating techniques is 3.8 billion years for the rocks from the Isua series in Greenland.* In either case, the surprising implication is that we may almost say that life has always existed on earth. Before 3.98 billion years ago (from 4.6 to 3.98 billion years), the earth was probably too hot to support life.[14] Then life appeared about 3.81 billion years ago. That is, only 0.170 billion (170 million) years were available for the abiotic emergence of life. Indeed, according to Brooks and Shaw, this amount of time for abiogenic synthesis of essential precursors, let alone chemical evolution, is "very small."[15] The discovery of microfossils has confirmed this conclusion. As a result, the thinking of scientists has undergone dramatic change. In the words of Miller, "If the origin of life took only 10^6 years [0.001

*Recently a zircon from the Australian Shield area has been dated at 4.2 billion years old. *Chem. Eng. News*, August 22, 1983, p. 20; *Science News*, June 18, 1983, p. 389.

billion], I would not be surprised."[16] Other scientists suspect a period of 10^7 to 10^8 years or less following the time after the earth cooled. For instance, "If higher surface temperatures persisted until 4000 Ma [4 billion years] ago, then life probably originated about 3900 Ma ago."[17] The search is underway for mechanisms that could account for the "geologically instantaneous" origin of life.

The Composition of Earth's Primitive Atmosphere

During the past several years, space probes have examined the atmospheres of several planets in our solar system. These probes have included investigations of the following planets:

1. Mars (Viking Missions).
2. Venus (Pioneer and Venera Missions).
3. Jupiter (Voyager Missions).
4. Saturn (Voyager Missions).

The data collected by these space probes have resulted in the re-examination of scientific theories concerning the formation of planets and their atmospheres. For example, the Pioneer Venus argon-neon measurements provided much-needed constraints on models of how modern atmospheres were generated. James B. Pollock of NASA-Ames has suggested three logical possibilities:[18]

1. The Primary Atmosphere Hypothesis
The gases in the modern atmosphere could be residuals from the pre-solar nebula. But if this were the case, the argon-neon ratios on Venus, earth, and Mars would be quite similiar to the original ratio in the nebula and the contemporary ratio on the sun. However, the ratios of these planetary atmospheres are very different from that of the sun.

2. The External Source Hypothesis
The gases could have been brought in on volatile-rich comets and asteroids in the post-T-tauri wind era while the planets were sweeping up the last pieces of matter from the solar system. These comets and asteroids must have bombarded all the inner planets at about the same rate; therefore, we would expect the planets to contain similiar concentrations of the rare gases. However, this is not the case.

3. The Grain Accretion Hypothesis

The modern planetary atmosphere could have resulted from outgassing of volatiles trapped in the original rocks.

According to Pollock, the last hypothesis is the only one not contradicted by the data. The term "grain accretion" is used because grains of material containing potential volatiles were accumulated into planetesimals that subsequently accreted to form planets.[19] Later, as a result of internal heating, volatiles reached the surface. Since the original volatile atmosphere of the earth escaped its gravitational field during accretion, the earth's primitive atmosphere was in fact a secondary atmosphere that resulted from gases issuing forth from the interior of the earth by means of volcanoes or by means of diffusion through the mantle. This secondary atmosphere theory has been the most accepted theory for over a decade, even with the influx of new information from Venus, Mars, and other planets.

Despite wide acceptance of the outgassing model, other sources of gases have been suggested to supplement it. For example, interstellar cloud material could be responsible for much of the neon in the earth's atmosphere.[20] Comets also may have supplied some of the volatiles.[21] Oro has estimated that 1,000 meteorites may have accounted for the volatiles on the earth.[22]

Various Models for the Earth's Primitive Atmosphere

In contrast to the wide acceptance enjoyed by the outgassing model for the formation of the atmosphere, opinions about the composition of the atmosphere have varied greatly over the years. Some examples of compositions postulated over the past 30 years follow:

The CO_2-H_2O Atmosphere. Assuming the volcanic exhalations to be the same on the primitive earth as today, the primitive atmosphere would be composed of carbon dioxide and water vapor with minor amounts of H_2S, SO_2, and N_2. This view was expressed by Fox and Dose,[23] Revelle,[24] Abelson,[25] and Brooks and Shaw.[26]

The CH_4 - NH_3 - H_2O Atmosphere. An opposing view was held by Oparin,[27] Urey,[28] and Miller and Urey.[29] These scientists reason that a small but significant level of H_2 remained in the atmosphere of the forming earth so that at least 10^{-3} atmosphere was present (there is about 10^{-6} atmosphere of H_2 today). The hydrogen would have reacted with any carbon, nitrogen, or oxygen present to form an atmosphere rich in methane (CH_4), ammonia (NH_3), and water (H_2O).

Of course, scientists of the first view disagreed with this conclusion, stating that the atmospheric H_2 level was insignificant and that there is no geologic evidence for a primitive atmosphere containing CH_4.[30]

The Three-Stage Atmosphere. A third view, held by Holland,[31] was really the synthesis of the first two views. Holland disagreed with the basic assumption of the first view, stating that the composition of gaseous mixtures from volcanoes of the primitive earth *was not* similiar to that of present-day volcanic exhalations. This came from the hypothesis that primitive volcanic exhalations, unlike their present counterparts, were in equilibrium with hot molten rock containing large amounts of elemental iron. This led to a first stage rich in methane (CH_4) followed closely by a second stage rich in N_2. The present-day atmosphere is the third stage.

The CO_2 - N_2 Atmosphere. Walker[32] has done an extensive study on the evolution of the atmosphere and concludes that the primitive atmosphere contained H_2O, CO_2, N_2, and 1% H_2. The 1% H_2 was emitted from volcanoes, and therefore he assumed that the volcanic source of hydrogen gas was larger in the past than today. Large quantities of the CO_2 emitted formed carbonates in oceans while large amounts of the H_2O condensed.

According to this view, the prebiological atmosphere contained no large amounts of reduced gases like methane and ammonia.[33] Recent photochemical calculations indicate that a heavily reducing atmosphere of methane and ammonia was extremely short-lived, if such a prebiological atmosphere existed at all.[34] The conclusion that the primitive atmosphere had little or no methane or ammonia has also won agreement from Holland.[35]

The notion that the primitive atmosphere was not highly reducing is a dramatic change from the previously held hypothesis. Various reports have elaborated on this shift in theories. For example:

> Now, for the first time in 30 years, the widely accepted recipe for primordial soup is changing from one rich in hydrogen—composed primarily of methane (CH_4) and ammonia (NH_3)—to a hydrogen-poor atmosphere similiar to today's *sans* the oxygen.[36]

> No geological or geochemical evidence collected in the last 30 years favors a strongly reducing primitive atmosphere.... Only the success of the laboratory experiments recommends it.[37]

> Scientists are having to rethink some of their assumptions. Chemists liked the old reducing atmosphere, for it was conducive to evolutionary experiments.[38]

Sherwood Chang of NASA-Ames Research Center has observed that prebiotic simulation experiments using a neutral atmosphere of water, nitrogen, and carbon dioxide produce only such chemicals as ammonia and nitric acid. [39] However, Joseph Pinto of the Goddard Institute for Space Studies synthesized formaldehyde in a primitive atmosphere poor in hydrogen.[40] Other simulation experiments using hydrogen-poor atmospheres have also produced abiotic organic molecules.[41] As reported in 1951, Melvin Calvin of the University of California at Berkeley synthesized organic compounds by irradiating a mixture of water and carbon dioxide with a beam of alpha particles.[42]

Oxygen Content of the Early Earth and its Atmosphere

All Models Exclude O_2

Models for the primitive atmosphere are many and diverse. Each scientist uses one of these atmospheric models to demonstrate that the chemical building blocks of life could be formed under the chosen conditions. However, an interesting pattern emerges from these experimental studies which suggests that, within limits, the syntheses of amino acids and other essential organic molecules are unexpectedly independent of the specific details of the experimental conditions. As discussed in Chapter 3, reactions that begin with an atmosphere of CH_4 and NH_3 or of CO_2 and N_2 as the carbon and nitrogen sources respectively are likely to result in similar products. Therefore, while a detailed evaluation of the primitive atmosphere is fascinating, it may not be necessary except for one point. That point, central to the theory of chemical evolution, is that the primitive atmosphere could not contain any but the smallest amount of free (molecular) oxygen (O_2).

It is necessary to exclude oxygen for two reasons. First, all organic compounds (such as the essential precursor chemicals or basic building blocks that must have accumulated for chemical evolution to proceed) are decomposed rather quickly in the presence of oxygen.

Second, if even trace quantities of molecular oxygen were present, organic molecules could not be formed at all. In the words of Shklovskii and Sagan, "As soon as the net [laboratory] conditions become oxidizing, the organic syntheses effectively turn off."[43] All the simulation experiments reviewed in Chapter 3 are largely inhibited by oxygen. None of the essential molecules of life, e.g. amino acids, could even be formed under oxidizing conditions, and if by some chance they were, they would decompose quickly. Chemical evolution would be impossible. This point is also made by Fox and Dose,[44] who list six reasons the primordial atmosphere contained no significant amount of oxygen. Two of their reasons are worthy of note: (1) "laboratory experiments show that chemical evolution ...would be largely inhibited by oxygen,"[45] and (2) "organic compounds that...have accumulated on the surface of the earth in the course of chemical evolution, are not stable over geologic time in the presence of oxygen."[46]

Fox and Dose hold the conviction that chemical evolution did occur, and list these points along with others as evidence for a reducing atmosphere. They reason that since chemical evolution requires it, free oxygen in the primitive atmosphere must have been negligible.

Fox and Dose are not the only ones who reason in this way. Walker[47] also concludes that the "strongest evidence" for an atmosphere without oxygen is that we know chemical evolution took place. While this may be an appropriate consideration for framing an hypothesis, it does not properly constitute evidence for the hypothesis.

We will discount this "strongest" evidence for an anoxic (no free oxygen) atmosphere since it is based on a circular argument. Such logic is hardly scientific, and simply assumes as true the hypothesis to be established. Without assuming in advance a reducing atmosphere, we will examine evidence concerning the oxygen content of the early earth's atmosphere. We will first consider sources of oxygen, and then examine mineralogical evidence during the time period over which oxygen has been present in the atmosphere. This, in turn, will help us determine when and for how long the earth's atmosphere was void of oxygen.

Sources of Free Oxygen for the Earth's Atmosphere

There are at least three possible sources of free oxygen in the earth's early atmosphere: volcanic exhalations (and comets/meteorites), photodissociation of H_2O, and the oxygen generating photosynthesis which is associated with living organisms. We will consider each of these sources in terms of the amount of oxygen produced and its probable date of appearance in geological history.

Volcanic Exhalation as a Possible Source of Free Oxygen. It has previously been suggested that the earth's atmosphere was produced by volcanic eruptions which might have included free oxygen (O_2) among the various gases. Gases from volcanic eruptions today contain mainly CO_2, H_2O, and minor amounts of H_2S, SO_2, and N_2, but no free oxygen. Given the high temperatures in volcanoes and the highly reactive nature of oxygen, this is not surprising. At elevated temperatures (600-800°C), oxygen would react with minerals in the earth resulting in nonoxidizing gases. We are thus left with neither a theoretical nor an experimental basis for expecting the early volcanic emissions to have supplied any significant amount of free oxygen to the primitive atmosphere.[48]

Photodissociation of Water as a Possible Source of Free Oxygen. Another possible source of free oxygen to the early atmosphere is the photodissociation of water in the atmosphere due to ultraviolet light or

$$2H_2O + (h\nu) \text{ ultraviolet light energy} = 2H_2 + O_2.$$

Since the 1960s, estimates of the amount of free oxygen in the prebiological atmosphere from photodissociation of water have ranged from 10^{-15} of present atmospheric level (PAL) to 0.25 PAL. The various estimates are provided in table 5-1 and summarized briefly below. It will be helpful to keep in mind that table 5-1 includes some entries listed as PAL and others as mixing ratio, where 1.0 PAL of oxygen is equivalent to a 0.21 mixing ratio (MR).

Berkner and Marshall[49] were the first to provide quantitative estimates of the concentration of oxygen in the early atmosphere resulting from photodissociation of water vapor. They concluded that concentrations of 10^{-3} PAL would have resulted.

Table 5-1
Estimates of Oxygen in the Early Atmosphere Due to Photodissociation

Author	*Year*	*Concentration*	*Footnote*	*Reference*
Berkner and Marshall	1965	10^{-3} PAL	1	49
Brinkmann	1969	0.25 PAL		50
Walker	1978	10^{-13} MR		33, 56
Kasting et al.	1979	10^{-12} PAL	2	56
VanderWood and Thiemens	1980	10^{-10} MR		56
Kasting and Walker	1981	10^{-8} PAL	3	56
Carver	1981	10^{-1} PAL		54
Levine	1982	10^{-15} PAL	4	34
Canuto et al.	1982	10^{-11} to 10^{-9} MR		57

(1) 1.0 PAL O_2 = 0.21 Mixing Ratio (MR).
(2) 10^{-12} PAL at surface, increases to 10^{-5} PAL at 60 Km.
(3) 10^{-14} when reaction of O_2 and H_2 included.
(4) 10^{-15} PAL at surface, increases to 10^{-5} PAL at 50 km altitude (a strong altitude dependence).

Brinkmann[50] calculated the amount of O_2 generated from photodissociation and consumed in oxidation of rock, etc. He concluded that a minimum of 25% of the present level (0.25 PAL) of oxygen existed over 99% of geologic time. Therefore, he reasoned, "It does not seem that early [chemical] evolution could have proceeded in such an atmosphere."[51] Proponents of a neutral or reducing early atmosphere do not agree that such high O_2 levels resulted from photodissociation of H_2O. For example, Walker[52] contends that Brinkmann erred in assuming that the rate of hydrogen escape from the earth is equal to the rate of photolysis of water. Walker, however, must assume that the volcanic source of hydrogen was considerably larger than the amount of hydrogen escaping into space after water was photolyzed. For this to have been true, volcanic sources of gases must have been much larger in the past than they are today. Van Valen[53] also objected to Brinkmann's study but failed to produce an alternative answer, offering only that there are serious and unresolved problems concerning the buildup of oxygen in the atmosphere.

Because of the importance of the question, Carver[54] recalculated the quantity of oxygen produced by photodissociation in Precambrian times using a larger water vapor mixing ratio than did previous studies. This study supports a warmer and more humid climate in the Precambrian. It also suggests that the free oxygen concentration could have reached 10% of the present level (0.1 PAL). If the surface oxidation rates were substantially greater in Precambrian times than at present, oxygen levels were probably 0.01 to 0.1 PAL.

Holland[55] has stated that a few percent of the present atmospheric level of oxygen was certainly present by 2.9×10^9 years ago. However, as shown in table 5-1 the estimates cover too broad a range to draw definite conclusions. Additional estimates not discussed here have been included in table 5-1 to illustrate the uncertainty in oxygen estimates.[56] The only trend in the recent literature is the suggestion of far more oxygen in the early atmosphere than anyone imagined. A significant part of this trend is due to measurements which suggest that stars resembling the sun at a few million years of age emit up to 10^4 times more UV light than the present sun.[57] This increase in UV could increase the O_2 surface mixing ratio by a factor of 10^4 to 10^6 over the standard value of 10^{-15}, thus affecting all the oxygen level estimates.[58]

Support for large estimates of O_2 is found in data from Apollo 16—data which suggest that a large amount of free oxygen does result from upper atmosphere photodissociation of water vapor. The Apollo 16's ultraviolet camera/spectrograph revealed a massive cloud of atomic hydrogen enveloping the earth and extending outward some 40,000 miles. This hydrogen apparently resulted from the photodissociation of water vapor. An early report of these results noted that this lends "substantial support" to "the theory that solar separation of water vapor provides our primary oxygen source" today and not photosynthesis as is usually supposed.[59]

George Carruthers,[60] principal investigator for the Apollo 16 camera/spectrograph experiment, has subsequently noted that the amount of oxygen due to photodissociation was originally overestimated. That is, photodissociation was not the primary source of oxygen as originally stated. (More details concerning the results of the measurements by Apollo 16 can be found in a report by Carruthers et al.)[61] Carruthers agrees with other workers that little free oxygen was present in the earth's primitive secondary atmosphere. However, without free oxygen (and therefore without ozone) solar ultraviolet radiation could penetrate to much lower water-rich layers of the atmosphere than is the case at present. Therefore, the water dissociation rate could have been much higher and the production rate of oxygen would have been considerably greater than at present. Thus, one may reasonably infer that the water vapor photodissociation process could have provided a sufficient amount of oxygen in the primitive atmosphere (perhaps as much as 1% of the atmosphere or 0.05 PAL) so that an ozone layer could have formed. An effective ozone screen would have allowed living organisms to proliferate by reducing the adverse effects of the solar UV radiation penetrating to ground level.

When asked about oxygen destroying organic molecules, Carruthers acknowledged it would, but not as rapidly as present-day oxidation because oxygen would have been more dilute and would not have been assisted by bacterial decay.[62] However, considering the long time postulated for chemical evolution to occur, even a small amount of oxygen would have been very detrimental. Most likely, if a small amount of O_2 were present, important precursor molecules would have been destroyed (oxidized) or their formation prevented in the first place.

Since living organisms and organic molecules need the protection from ultraviolet radiation provided by an ozone screen, yet the presence of oxygen prevents the development of such living systems and biological molecules, this would seem to constitute a catch-22 in the model. How much oxygen is required to produce the ozone screen and what maximum amount of oxygen can be tolerated in the synthesis of the molecular precursors to life? These two questions will be considered next.

Berkner and Marshall[63] were among the first scientists to evaluate the relationship of O_2 to O_3 as it pertains to chemical evolution. They suggested that when the O_2 concentration reached 10^{-2} PAL, the resulting concentration of O_3 was sufficient to restrict the penetration of lethal UV to a thin layer of the ocean. When the O_2 level reached 10^{-1} PAL, the O_3 concentration was sufficient to absorb all UV radiation less than 3000 Å. At these levels, life was able to migrate from the oceans to land masses for the first time. Since this initial evaluation by Berkner and Marshall, other scientists have investigated the origin and evolution of ozone.[64]

The suggestion has been made that very little atmospheric oxygen (possibly 10^{-3} PAL), is required to produce a biologically effective ozone screen. However, when several additional factors are taken into account it becomes apparent that perhaps as much as 0.1 PAL oxygen would have been required. Carver,[65] in reviewing the available data, concluded that a biologically effective ozone screen would be established once the oxygen content exceeds 0.01 PAL.

In summary, the development of an ozone screen apparently requires a higher oxygen concentration (0.01 to 0.1 PAL) than the original suggestion of 10^{-3} PAL. Whether such a free oxygen concentration developed by photodissociation of water alone, or eventually by the combined action of photodissociation and photosynthesis in algae, etc., is difficult to establish. It is not yet known at what rate free oxygen is removed by reaction with reducing gases such as methane or reduced minerals such as Fe_3O_4. In any case, it seems evident that free oxygen was being produced by photodissociation from earliest times and that this source of free oxygen would have continued until a significant free oxygen concentration developed allowing an ozone screen to form, filter the short wavelengths (i.e., < 3000 Å) of ultraviolet light, and effectively turn off this mechanism of oxygen production in the atmosphere beneath the ozone screen.

Because only low levels of oxygen are needed, the earth may have had an effective ozone screen since before life began. Such a prospect makes this area of research quite controversial.

Two consequences of an early ozone screen are:

1) the requirement that sources of energy other than UV light would need to be postulated for prebiotic synthesis of organic molecules, and
2) the necessity of alternative scenarios which would allow substantial synthesis of organic molecules and their subsequent protection in an oxidizing milieu.

Living Organisms as a Source of Free Oxygen. Since volcanic eruptions apparently would not supply free atmospheric oxygen and photodissociation would supply free oxygen only until an ozone layer developed (apparently between 0.01 and 0.1 PAL of oxygen), it is generally assumed that our present 21% of free atmospheric oxygen was and is the result of photosynthesis by living plants. This transition from the *assumed* anoxic conditions to our present 21% free oxygen is usually thought to have occurred about 1-2 billion years ago. Figures 5-1 and 5-2 illustrate estimates by several scientists of the increase in O_2 with time.

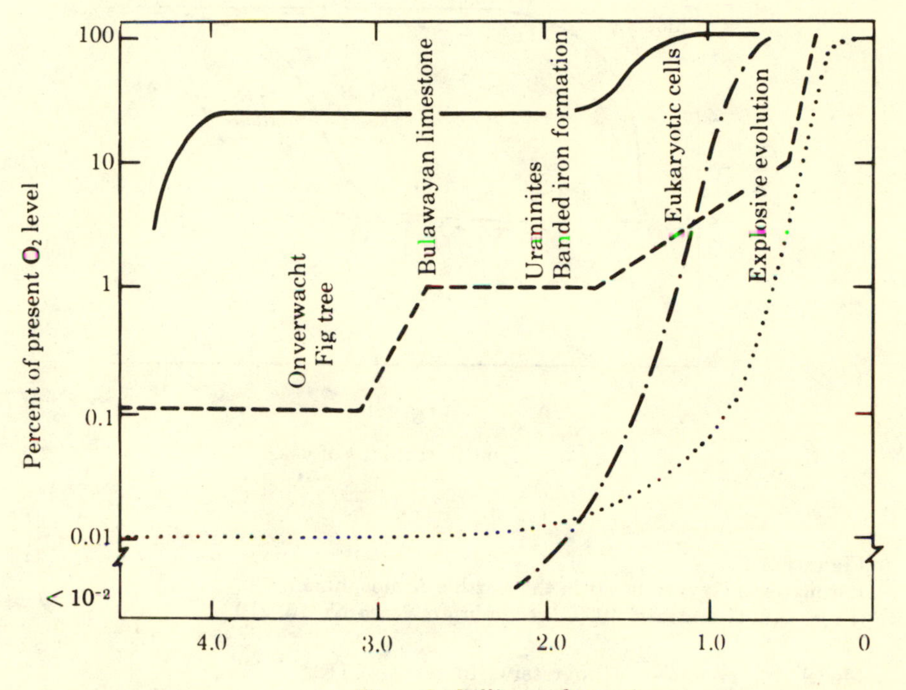

Figure 5-1.
Estimates of Oxygen Levels in the Earth's Atmosphere.
(From S.L. Miller and L.E. Orgel, 1974. *The Origins of Life on the Earth*, p. 52.)

....	**Berkner & Marshall**
— — — —	**Rutten**
—.—.—.	**Walker (derived from text)**
———	**Brinkmann (derived from text)**

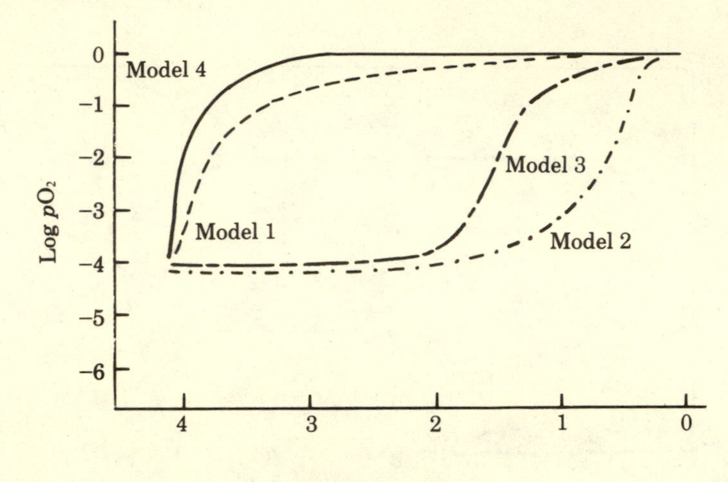

Figure 5-2.
Estimates of Oxygen levels in the Earth's Atmosphere.
(From D.E. Grandstaff, 1980. *Precambrian Research*, **13**, 21.)

Model 1:	-----------	Progressive Increase of O_2
Model 2:	—.—.—.	O_2 remained low until Devonian (Berkner and Marshall)
Model 3:	— - — -	O_2 remained low until red beds disappeared (Cloud)
Model 4:	————	O_2 increased rapidly to present level (Kimberly and Dimroth)

However, recent paleontological evidence suggests an advent of a more highly oxidizing atmosphere earlier than 1 to 2 billion years ago. At the beginning of this chapter, we discussed the age of the first life on earth. Some of these life forms would have produced oxygen. Still, the level of O_2 production remains in doubt. The organisms could have been anaerobic bacteria, in which case the atmosphere could have been anoxic. Walker[66] dates autotrophic organisms at 3.5 billion years ago, bacterial photosynthesis at 3 billion years ago, and the advent of green-plant photosynthesis at about 2.5 billion years ago. Thus oxygen-producing organisms (cyanobacteria/blue-green algae) certainly existed by 2.8×10^9 years ago and perhaps much earlier (probably 2.9-3.1 billion years ago). According to Schopf[67] these organisms would have produced fluctuating levels of free oxygen. At first, the oxygen would have been consumed by exposed reduced-mineral species (mainly ferrous iron). Then the quantity of oxygen would have varied depending on the exposure of more reduced minerals, the amount of volcanic emissions, etc. until the concentration reached fairly constant levels about 2 billion years ago. Until recently, however, most scientists thought that little oxygen existed before 2 billion years ago. Walker mused, "...it is hard to explain why oxygen pressures should have remained low for almost 2 billion years after the introduction of green plant photosynthesis."[68]

Based on the growing body of evidence, Walker has concluded that oxygen-evolving photosynthesis appeared prior to 3.8 billion years ago and that the lifetime of the prebiological atmosphere must have been "quite short in geologic terms."[69]

From the available data on isotopic sulfur composition of Precambrian minerals, Chukhrov et al. have concluded "...the existence of sulfate-reducing organisms and the presence of substantial amounts of oxygen in the terrestrial atmosphere 3000 m.y. ago or earlier."[70] Likewise, from carbon isotope studies, Eichmann and Schidlowski have shown that more than "3 billion years ago photosynthesis [had] produced already a large fraction of all the oxygen ever released and now fixed primarily in Fe_2O_3 and SO_4^{2-} with only 5% present as free oxygen in the atmosphere."[71] The data of Schidlowski, et al.[72] also show no secular change in the isotopic composition of carbonates dating back more than 3 billion years ago. Even more recently Schidlowski has indicated that "the constancy of the isotopic fractionation observed between reduced and oxidized carbon throughout the record is best interpreted as the signature of biological activity during the past 3.5×10^9 yr. (or possibly 3.8×10^9 yr.)."[73] Broecker[74] considers such constancy of $^{13}C/^{12}C$ ratios in Phanerozoic (younger than 0.6 billion years) marine carbonates as indicative that the oxygen content must have been comparable to its present value. If this principle is valid for Phanerozoic carbonates, it should also be valid for carbonates 3 billion years ago. That is, we must conclude that the present level of oxygen also existed 3 billion years ago. Based on Schidlowski's data, other scientists have concluded that 80% of the present levels of oxygen have existed for the past 3.0 billion years.[75]

Oxygen-producing organisms probably formed very old limestone deposits (e.g., Bulawayan, 2.7-3.0 billion years) in the same manner as do the present-day limestone-depositing algae. Judging from the amount of limestone in ancient deposits, significant levels of O_2 would have been present. However, Rutten[76] disagrees with this conclusion and contends that since the O_2 concentration 2.7 billion years ago was only 1% of the present level, the metabolism of limestone-depositing organisms must have been different in the past from present algae. But we must ask, why change the metabolism of the algae? Surely the desire for a prebiotic earth without free oxygen is not a compelling reason. It would have been just as easy (or easier?) to adjust the O_2 level to account for the limestone.

In summarizing this section on sources of free atmospheric oxygen, the most likely scenario is as follows. The early secondary atmosphere contained mainly N_2, H_2O, and CO_2. Photodissociation then produced an indeterminate free-oxygen concentration which was later supplemented by photosynthesis. Once the oxygen level reached a concentration of 0.01 to 0.1 PAL (by photodissociation alone or in combination with photosynthesis), an effective ozone layer formed and photodissociation ceased in the lower atmosphere. The remaining increase in oxygen concentration to present levels occurred by photosynthesis alone. Recent paleontological data combined with occurrence of living organisms 3.5 billion years ago indicate that these increases in oxygen levels may have occurred very early in geological history (over 3 billion years ago).

This scenario raises two very significant questions. First, what free oxygen concentration level was produced by photodissociation acting alone before the origin of life? And second, would this level of free oxygen adversely affect the formation or continuance of organic biomonomers? We have already addressed the first question and found that current estimates of O_2 in the early atmosphere resulting from photodissociation range from 10^{-15} PAL to 10^{-1} PAL. Levine states, "This is a wide range, even for studies of the paleoatmosphere. Additional research in this area is indicated."[77] The second question is equally difficult to answer in a precise manner. Only qualitative statements have been made. For example:

> Even at low levels of O_2, there is a slow oxidation of most organic compounds, and the rate is greatly enhanced in the presence of ultraviolet light. These and related arguments are so compelling that it does not seem possible that organic compounds remained in the primitive ocean for any length of time after O_2 entered the earth's atmosphere.[78]

We can only say, based on current models for ozone formation, that the upper limit of free oxygen concentration resulting from photodissociation alone would be 0.01 to 0.1 PAL. As indicated, there is considerable controversy concerning whether this upper limit of oxygen concentration could have been reached by photodissociation alone. Current estimates of 10^{-15} PAL surely are too low for production of an ozone screen, while 10^{-1} PAL is the upper limit itself. One thing is clear: if further research confirms that photodissociation alone could have produced a biologically effective ozone screen, a second problem is inescapable. Enough oxygen would then have been present in the early atmosphere to effectively shut off any production and/or accumulation of biomonomers, thus preventing chemical evolution.

Mineral Evidence Pertinent to Defining Free Oxygen Content in the Atmosphere during Various Stages of Geological History

The results from atmospheric physics, while not conclusive about the oxidation state of the early atmosphere, do at least leave open the possibility the early earth was oxidizing. This possibility is in conflict with the usual picture of the early earth as reducing. Therefore, we shall re-examine the data and usual agruments supporting the notion of a reducing early earth and atmosphere.

The interpretation of the mineral evidence pertinent to atmospheric free oxygen in geologic history depends on the oxidation states of elements in mineral deposits that were formed during the various geological periods. For example, in the reaction

$$PbS + 2O_2 \longrightarrow PbSO_4$$

at 25°C, the equilibrium pressure of O_2 for the oxidation of PbS to $PbSO_4$ is 10^{-63} atm. This equilibrium pressure is so small that if any oxygen were present PbS would be converted to $PbSO_4$. Therefore, if rocks can be found to contain PbS versus $PbSO_4$, it would seem reasonable to conclude they formed in an anoxic environment. Likewise, if $PbSO_4$ is more abundant than PbS, oxygen may be inferred to have been present at its formation. It is instructive to note that other minerals show a similar relationship:

Reduced Form (formed under anoxic conditions)		*Oxidized Form* (formed under oxygenic conditions)
Fe_3O_4	(Magnetite)	Fe_2O_3 (Hematite)
$UO_2 - U_3O_8$	(Uraninite)	UO_3
PbS	(Galena)	$PbSO_4$
ZnS	(Wurtzite)	$ZnSO_4$ or $ZnSO_4 \cdot 7H_20$
$Fe_{1-x}S$	(Pyrrhotite)	$FeSO_4 \cdot 7H_2O$

The thermodynamic data indicate that the equilibrium oxygen pressures for the oxidation of the sulfides (PbS, ZnS, and FeS) to the corresponding sulfates (SO_4^{2-}) are lower than the equilibrium pressure for conversion of

$U_3O_8 - UO_2$ to UO_3 ($pO_2 = 10^{-21} - 10^{-23}$ atm.)

The equilibrium O_2 pressure for the conversion of Fe_3O_4 to Fe_2O_3 ($pO_2 = 10^{-72}$ atm.) is even less than the values for sulfide oxidations. A comprehensive review of the various elements and the oxidation states used in this type of study has been summarized by Rutten.[79]

Basic Assumptions. Interpretation of mineral data involves two basic questions. First, how long does it take for a given mineral to oxidize? And second, how long was the mineral in question exposed to the atmosphere during formation or exposed thereafter during transportation and deposition? It is usually assumed that a reduced or only partially oxidized mineral was formed when the atmosphere was anoxic, but this is not necessarily the case. We must also consider the rate of the reaction (kinetics). The predictions of equilibrium thermodynamic data are only significant if given enough time. If the mineral is not in contact with the atmosphere or water saturated with the atmospheric gases for sufficient time during transportation and deposition, it will not come to equilibrium. Since some of these reactions are very sluggish at ambient temperatures, the presence of a reduced mineral or absence of a fully oxidized mineral does not necessarily mean that the atmosphere was anoxic. Several examples are offered by way of illustration in the following paragraphs.

Specific Examples of Mineral Assemblages: Iron and Uranium Oxides

1. Iron Oxides

It is by no means unequivocal that iron oxides indicate ancient O_2 levels. This is best demonstrated by examining the stability fields of different iron minerals under varying natural conditions of pH and oxidation/reduction potential. When the O_2 level is changed from the present level to 0.01 PAL, the stability fields change very little (see figure 5-3). That is, the stability and depositional conditions of the iron oxides are hardly affected. Rutten concluded, "It follows that arguments in favor of an anoxygenic atmosphere cannot be based on the equilibria of mineral reactions,...but on their kinetics,"[80] or the rate at which oxidation occurs.

According to Fox and Dose[81] no agreement has been reached concerning the equilibrium between FeO, Fe_3O_4, and Fe_2O_3 as a function of O_2 level. Holland[82] points out that Fe_2O_3 would be stable under extremely low O_2 levels, which explains its existence in sediments greater than 2.5 billion years old when the atmosphere was thought to have contained no oxygen. But other geologists use the occurrence of Fe_2O_3 to indicate significant levels of O_2 in the primitive atmosphere. Davidson[83] states that such immense hematite (Fe_2O_3) deposits (as far back as 3.4 billion years ago) are only compatible with the presence of free oxygen in surface waters at this very early date. The fact that all oxidation states of iron, from FeO to Fe_2O_3 to FeS_2, have been found in sediments of all ages probably indicates that local conditions and not the overall conditions determine which particular mineral is present. For example, as recently as 0.4 to 0.5 billion years ago (when O_2 was at its present level), reduced minerals were being deposited in oxygen-free waters (a local anoxic environment) much like the Indian Ocean today, which has practically no free oxygen below 150 meters. By looking at these deposits, one would erroneously conclude that the atmosphere was anoxic at that time. Such data led Krecji-Graf to conclude that geological evidence cannot be used to make general deductions concerning the earth's atmosphere.[84] Another explanation of the observed variation of oxidation states of iron is that the levels of oxygen fluctuated in the ancient atmosphere. Schopf indicates that such conditions probably existed over 3.0 billion years ago.[85]

Despite the inconclusive nature of oxygen levels and iron formations, the customary interpretation has been that red beds (Fe_2O_3)

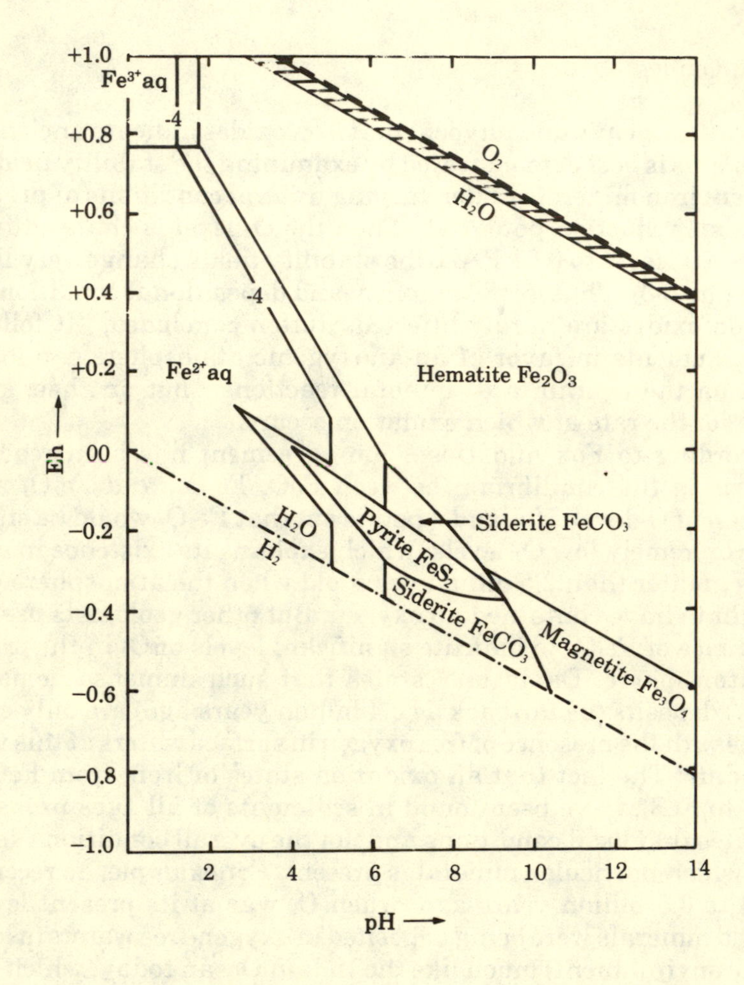

Figure 5-3.
Iron Compound Stability Fields.

Stability fields of iron compounds in water at 25°C and 1 atm. total pressure, when the activity of total dissolved sulfur and carbonate is 10^{-6} and 1.0, respectively. The barred area indicates the change in stability fields when O_2 is decreased from 1.0 PAL (upper line) to 0.01 PAL (lower line). It hardly affects the stability field of hematite. (Redrawn from Rutten, 1971. *The Origin of Life by Natural Causes*. New York: Elsevier, p. 82.)

provide the best indication of the first appearance of oxygen.[86] Walker disagrees, however, stating, "The presence of banded iron formation in the Isua rocks of West Greenland therefore implies that oxygen-evolving photosynthesis appeared on earth prior to 3.8 billion years ago."[87] Walker's reasoning assumes that many metabolic processes capable of affecting the atmosphere (e.g., fermentation, bacterial photosynthesis, and sulfate reduction) must have originated before oxygen-evolving photosynthesis. Therefore, the lifetime of the prebiological atmosphere of nitrogen, carbon dioxide, and water vapor must have been quite short in geological terms.

2. Uranium Oxides

A somewhat clearer picture emerges from UO_2 - UO_3 deposits of the Dominion Reef and Witwatersrand system in South Africa. The mineral deposits contain uraninite (UO_2), galena (PbS), pyrite (FeS_2), and gold. The deposits are all sedimentary. The minerals were derived from weathering a granite source rock and carried by high-energy (steep, fast-flowing) rivers to a lower-energy (flat, slow) fan-delta system where the minerals were deposited. This is evidenced by the well-rounded, coarse-silt-sized (0.0655mm) uraninite grains in the deposit. This type of deposit is called a detrital or placer deposit and the environment in which it was deposited is called a fluvial fan-delta or a braided alluvial plain. The minerals were definitely in contact with the atmosphere as they were weathered and deposited, some 2.5-2.75 billion years ago. Because the reduced forms of the minerals are present, it is usually concluded that the deposits were formed under an anoxic environment. However, as Miller and Orgel point out, "...these minerals may have been deposited under local reducing conditions, or failed to have reached equilibrium with the atmosphere at the time they were laid down."[88] Most geologists, however, would readily conclude that the minerals were in equilibrium due to the river transport as detailed above. However, this too is a matter of kinetics. If the minerals were transported and deposited very rapidly, for example, they may not have had time to reach equilibrium with the atmosphere. If this were the case, the reduced UO_2 would still be deposited in the presence of significant levels of O_2. But rapid deposition may not have occurred given that the individual mineral grains are well-rounded and sorted.

Another possibility is that these deposits were transported during glacial periods. The very cold environment would lower the rate of reaction of UO_2 with O_2. Therefore, UO_2 would be deposited in the presence of O_2. Some evidence exists of glaciers in South Africa 2.5 billion years ago, and present-day evidence indicates that UO_2 deposits are now being formed in cold environments. In fact there is evidence that detrital uraninite exists in the present-day Indus River of Pakistan.[89] This further illustrates the fact that the rates of reactions must be known before definite conclusions can be made.

Trow has proposed a mechanism for deposition of the Witwatersrand and Elliot Lake uranium deposits in an oxygenated atmosphere during glacial, CO_2-impoverished episodes. He states that "apparently an anoxic atmosphere did not exist at these times [2.25-2.5 billion years ago]."[90]

We agree with Walker[91] that the evidence for an anoxic atmosphere provided by the detrital uraninite and pyrite in the Witwatersrand is not strong. This is based upon work by Holland[92] that shows that an upper limit of about 1% of the oxygen-mixing ratio is consistent with the existence of detrital uraninite. Also, according to Muir,[93] detrital pyrite (a reduced mineral) is common even today. In summarizing the various contributions at the U.S. Geological Survey Quartz-Pebble Workshop, Skinner[94] stated that current theories on atmospheric control for such ores as the Witwatersrand are not well established. He further remarked that the current thinking is not correct and the absence of atmospheric oxygen cannot be counted upon with certainty to explain uraniferous quartz-pebble conglomerates. He suggested a more neutral atmosphere as an alternative to either a reducing or oxidizing atmosphere.

Much of the ambiguity about mineral assemblages has been resolved by D.E. Grandstaff,[95] who made a kinetic analysis of the oxidation of U^{4+} to U^{6+}. Uraninite (UO_2-U^{4+}) is thermodynamically unstable at oxygen pressures greater than approximately 10^{-21} atmospheres. Yet Grandstaff's kinetic analysis indicates that uraninite may have survived without being oxidized at oxygen pressures as high as 0.01 PAL. Thus deposition of uraniferous conglomerates

> ...does not require an essentially anoxic atmosphere as previously proposed, but may have occurred under an atmosphere containing small amounts of oxygen consistent with photodissociation of water vapor and limited aerobic photosynthesis.[96]

The important conclusion from Grandstaff's kinetic analysis is that the formation of a reduced mineral such as UO_2 or Fe_3O_4 need not have required the absence of free oxygen in the atmosphere at the time the mineral was formed. Thus, traditional arguments for a reducing atmosphere based on reduced minerals are unconvincing. At least a mildly oxidizing atmosphere of up to 0.01 PAL is possible without oxidizing U^{4+}. It has long been known that the proper understanding of a thermodynamically favorable reaction is simply a reaction that is permitted. It need not occur. Only by kinetic analysis can details be obtained of whether a reaction occurred, and at what rate.

Summary of Mineral Data. We have examined in detail the evidence from uranium and iron minerals concerning the existence of a reducing primitive atmosphere. Because of the uncertainty in the kinetics of oxidation of these minerals, it is difficult to conclude with confidence that there has ever been a time when the earth's atmosphere was devoid of free oxygen. Erich Dimroth and Michael Kimberley have evaluated minerals besides uranium and iron, and have drawn a similar conclusion:

> In general, we find no evidence in the sedimentary distribution of carbon, sulfur, uranium, or iron that an oxygen-free atmosphere has existed at any time during the span of geological history recorded in well-preserved sedimentary rock.[97]

Chapter Summary and Conclusions

Three relevant questions have been considered in this chapter. First, we considered the time available for chemical evolution. It was determined on the basis of evidence from molecular fossils and microfossils that the origin of life occurred almost instantaneously (geologically speaking), just after the earth's crust cooled and stabilized about 4.0 billion years ago. This leaves little more than 100 million years (if that) for any chemical evolution to occur. Second, the early atmosphere of the earth was examined and found not to be the strongly reducing atmosphere popularized for the past thirty years. Instead, the consensus of scientists about the early atmosphere is shifting. At the time of this writing, there is wide agreement in adopting a more neutral primitive atmosphere consisting of CO_2, N_2, H_2O, and perhaps 1% H_2. There is a current controversy concerning whether the early earth and its atmosphere might actually have been oxidizing. Third, we examined the important question of the oxygen content of the early earth.

Three lines of evidence have been evaluated that indicate the existence of free oxygen in the earth's primitive atmosphere: (1) data showing oxygen-producing life forms in rocks older than 3.5 x 10^9 years, (2) data showing oxidized mineral species in rocks older than 3.5 x 10^9 years, and (3) calculations indicating that up to 0.1 PAL of O_2 could have been produced by photodissociation of water. Although no precise conclusions can be made concerning the levels of oxygen in the earth's early atmosphere, these results are quite suggestive.

The accumulating evidence for an oxygenic early earth and atmosphere heightens the mystery of life's origin. If this type of evidence continues to accumulate, chemical evolution theories may have to appeal to the random occurrence of fluctuating or localized reducing environments on the primitive earth. Such micro-environments could have been present (as shown by reduced minerals), but were they suitable or maintained long enough for the formation of life? The odds of finding such a suitable niche on the primitive earth for a sufficient length of time are extremely small.

The monomer experiments reviewed in Chapter 3 largely assumed a strongly reducing atmosphere. These experiments covered the period from Miller's classic experiment reported in 1953 to the mid-1970s. In fact, one can mark the shift to a less-reducing atmosphere with the Viking Mission to Mars. Although, as Chapter 5 has shown, considerable evidence of an oxidizing early earth was available before 1976, the discovery of an oxidizing Mars void of life served to focus attention on the question of the oxygen history of earth.

As might be expected then, primitive atmosphere experiments will need to be reassessed in the light of evidence that the early earth and its atmosphere were probably less reducing than first suspected, and possibly even oxidizing. There are signs that this important process of re-doing experiments with more plausible atmospheres is underway. A few experiments using more neutral to mildly oxidizing atmospheres were mentioned earlier in this chapter. These experiments have generally yielded products in smaller quantities and less diversity than comparable experiments under more reducing conditions. However, there seems to be no less optimism regarding the prospects that chemical evolution was a near-certain occurrence on this planet.

References

1. R.E. Dickerson, 1978. *Sci. Am.* **239**, 70.
2. "How Did Life Begin?" Aug. 6, 1979, *Newsweek*, p. 77.
3. Dickerson, *Sci. Am.*, p. 70.
4. G. Tilton and R. Steiger, 1965. *Science* **150**, 1805; Robert H. Dott, Jr., Roger L. Batten, and Randall D. Sale, 1981. *Evolution of the Earth*. New York: McGraw-Hill, 3rd Ed., p. 157.
5. S. Fox and K. Dose, 1972. *Molecular Evolution and the Origin of Life*. San Francisco: W.H. Freeman & Co., p. 286.
6. Ibid., p. 289.
7. J.W. Schopf and E.S. Barghoorn, 1967. *Science* **156**, 508.
8. H. Knoll and E.S. Barghoorn. 1977. *Science* **198**, 396.
9. H.D. Pflug and H. Jaeschke-Boyer. 1979. *Nature* **280**, 483; C. Ponnamperuma, Sept. 24, 1979. *Time*; C. Ponnamperuma, Sept. 10, 1979. American Chemical Society Meeting, Washington, D.C.
10. J.W. Schopf. June 30, 1980. *Newsweek*, p. 61.
11. D.R. Lowe. 1980. *Nature* **284**, 441; M.R. Walter, R. Buick, and J.S.R. Dunlop, 1980. *Nature* **284**, 443.
12. D. Bridgewater, J.H. Allaart, J.W. Schopf, C. Klein, M.R. Walter, E.S. Barghoorn, P. Strother, A.H. Knoll, and B.E. Gorman, 1981. *Nature* **289**, 51; Nigel Henbest, 1981. *New Scientist* **92**, 164.
13. J. Brooks and G. Shaw, 1973. *Origin and Development of Living Systems*. London and New York: Academic Press, p. 73.
14. Ibid., p. 78.
15. Ibid., p. 78.
16. S.L. Miller, 1982. In *Mineral Deposits and the Evolution of the Biosphere*, eds. H. Holland and M. Schidlowski. New York: Springer-Verlag, p. 157.
17. S. Awramik, P. Cloud, C. Curtis, R. Folinsbee, H. Holland, H. Jenkyns, J. Langridge, A. Lerman, S. Miller, A. Nissenbaum, J. Veizer, 1982. In *Mineral Deposits and the Evolution of the Biosphere*, eds. H. Holland and M. Schidlowski, p. 311.
18. M. Waldrop, Feb. 19, 1979. *Chem. Eng. News*, p. 26; J.B. Pollock and D.C. Black, 1979. *Science* **205**, 56.
19. P.H. Abelson, 1979. *Science* **205**, 1.
20. D.M. Butler, M.J. Newran, R.T. Tolbert, Jr., 1978. *Science* **201**, 522.
21. L. Garmon, 1981. *Science News* **119**, 72.
22. J. Oro, 1971. In *Proceedings of the Second Conference on Origins of Life*, ed. L. Margulis. Washington, D.C.: The Interdisciplinary Communication Assoc., Inc., p. 7.
23. Fox and Dose, *Molecular Evolution and the Origin of Life*, p. 40.
24. R.T. Revelle, 1965. *J. Marine Res.* **14**, 446.
25. P.H. Abelson, 1966. *Proc. Nat. Acad. Sci. U.S.* **55**, 1365.
26. Brooks and Shaw, *Origin and Development of Living Systems*, p. 77.

27. A.I. Oparin, 1938. *The Origin of Life* (trans. by S. Morgulis). New York: Macmillan.
28. H.C. Urey, 1952. *The Planets*. New Haven, Conn.: Yale University Press.
29. S.L. Miller, H.C. Urey, 1959. *Science* **130**, 245.
30. Fox and Dose, *Molecular Evolution and the Origin of Life*, p. 43.
31. H.D. Holland, 1962. In *Petrologic Studies*, eds. A.E.J. Engel, H.L. James, and B.F. Leonard, Geological Society of America, p. 447.
32. J.C.G. Walker, 1977. *Evolution of the Atmosphere*. New York: Macmillan, p. 210, 246.
33. J.C.G. Walker, 1978. *Pure Appl. Geophys.* **116**, 222.
34. J.S. Levine, 1982. *J. Mol. Evol.* **18**, 161.
35. H.D. Holland, Harvard University, Dept. of Geological Sciences, personal communication, June 24, 1983.
36. Garmon, *Science News*, p. 72.
37. R.A. Kerr, 1980. *Science* **210**, 42.
38. R.C. Cowen, April, 1981, *Technology Review*, p. 8.
39. Waldrop, *Chem. Eng. News*, p. 26; Pollock and Black, *Science*, p. 56.
40. J. Pinto, G.R. Gladstone, and Y.L. Yung, 1980. *Science* **210**, 183.
41. K. Kawamoto and M. Akabosh, 1982. *Origins of Life* **12**, 133; C. Folsome, A. Brittain, and M. Zelko, 1983. *Origins of Life* **13**, 49.
42. W.M. Garrison, D.C. Morrison, J.G. Hamilton, A.A. Benson, and M. Calvin, 1951. *Science* **114**, 416.
43. I.S. Shklovskii and C. Sagan, 1966. *Intelligent Life in the Universe*. New York: Dell, p. 231.
44. Fox and Dose, *Molecular Evolution and the Origin of Life*, p. 44-45.
45. Ibid., p. 44.
46. Ibid., p. 45.
47. Walker, *Evolution of the Atmosphere*, p. 224.
48. Kenneth M. Towe, 1978. *Nature* **274**, 657.
49. L.V. Berkner and L.C. Marshall, 1965. *J. Atmos. Sci.* **22**, 225.
50. R.T. Brinkmann, 1969. *J. Geophys. Res.* **74**, 5355.
51. Ibid., p. 5366.
52. Walker, *Evolution of the Atmosphere*, p. 224.
53. L. Van Valen, 1971. *Science* **171**, 439.
54. J.H. Carver, 1981. *Nature* **292**, p. 136.
55. H.D. Holland, Harvard University, Dept. of Geological Sciences, personal communication, June 24, 1983.
56. J.C.G. Walker, 1978. *Pure Appl. Geophys* **117**, 498; J.F. Kasting, S.C. Liu, and T.M. Donahue, 1979. *J. Geophys. Res.* **83**, 3097; T.B. Vander Wood and M.H. Thiemens, 1980. *J. Geophys. Res.* **85**, 1605; J.F. Kasting and J.C.G. Walker, 1981. *J. Geophys. Res.* **86**, 1147.
57. V.M. Canuto, J.S. Levine, T.R. Augustsson, and C.L. Imhoff, 1982. *Nature* **296**, 816.
58. Ibid., p. 820.
59. News Release #30-72-7 from the Naval Research Laboratory, Washington, D.C., 1972 and Preliminary Report Lunar Surface Ultraviolet Camera/Spectrograph Apollo 16 experiment S-201.
60. G.R. Carruthers, Naval Research Laboratory, personal communication, Feb. 20, 1979, and Sept. 28, 1981.
61. G.R. Carruthers, T. Page, R.R. Meier, 1976. *J. Geophys. Res.* **81**, 1665.

62. G.R. Carruthers, personal communication, Feb. 20, 1979.
63. Berkner and Marshall, *J. Atmos. Sci.*, p. 225.
64. J.S. Levine, *J. Mol. Evol.*, p. 161; Carver, *Nature*, p. 136; M.L. Ratner and J.C.G. Walker, 1972. *J. Atmos. Sci.* **29**, 803; A.J. Blake and J.H. Carver, 1977. *J. Atmos. Sci.* **34**, 720; E. Hesstvedt, S. Henriksen, and H. Hjartarson, 1974. *Geophysica Norvegica* **31**, 1; J.F. Kasting and T.M. Donahue, 1980. *J. Geophys. Res.* **85**, 3255.
65. Carver, *Nature*, p. 136-8.
66. Walker, *Evolution of the Atmosphere*, p. 263.
67. J.W. Schopf. U.C.L.A., Dept. of Earth and Planetary Sciences, personal communication, June 23, 1983.
68. Walker, *Evolution of the Atmosphere*, p. 266.
69. Walker, 1978. *Pure Appl. Geophys.* **116**, p. 230.
70. F.V. Chukhrov, V.I. Vinogrador, and L.P. Ermilova, 1970. *Mineral. Deposita* (Berl.) **5**, 209. Quote from page 220.
71. R. Eichmann, M. Schidlowski, 1975. *Trans. Am. Geophys. Union* **56**, 176.
72. M. Schidlowski, R. Eichmann, C.E. Junge, 1975. *Precambrian Res.* **2**, 1; M. Schidlowski, 1976. In *The Early History of the Earth*, ed. B. Windley. New York: Wiley and Sons, p. 525.
73. M. Schidlowski, 1982. In *Mineral Deposits and the Evolution of the Biosphere*. eds. H. Holland and M. Schidlowski, p. 103.
74. W.S. Broecker, 1970. *J. Geophys. Res.* **75**, 3553.
75. S. Awramik, et al., in *Mineral Deposits and the Evolution of the Biosphere*, p. 314.
76. M.G. Rutten, 1971. *The Origin of Life by Natural Causes*. New York: Elsevier Publishing Co., p. 319.
77. Levine, *J. Mol. Evol.*, p. 167.
78. S.L. Miller and L.E. Orgel, 1974. *The Origins of Life on the Earth*. Englewood Cliffs, New Jersey: Prentice-Hall, p. 119; S.L. Miller, in *Mineral Deposits and the Evolution of the Biosphere*, p. 160.
79. Rutten, *The Origin of Life by Natural Causes*, p. 253.
80. Ibid., p. 282.
81. Fox and Dose, *Molecular Evolution and the Origin of Life*, p. 44.
82. Holland, in *Petrologic Studies*, p. 447.
83. Charles F. Davidson, 1965. *Proc. Nat. Acad. Sci. USA* **53**, 1194.
84. Graf K. Krejci, cited in Fox and Dose, *Molecular Evolution and the Origin of Life*, p. 44.
85. J.W. Schopf, U.C.L.A. Dept. of Earth and Planetary Sciences, personal communication, June 23, 1983.
86. Walker, *Evolution of the Atmosphere*, p. 262.
87. J.C.G. Walker, 1978. *Pure Appl. Geophys.* **116**, 230.
88. Miller and Orgel, *The Origins of Life on Earth*, p. 50.
89. P.R. Simpson and J.F.W. Bowle, 1977. *Phil. Trans. R. Soc. Lond.* **A286**, 527. Referenced in K.M. Towe, 1978. *Nature* **274**, 657.
90. James Trow, March 16, 1978. "Uraniferous Quartz-Pebble Conglomerates and their Chemical Relation to CO_2 - Deficient Atmosphere Synchronous with Glaciations of Almost Any Age," Dept. of Geology, Michigan State University, East Lansing, Mich. (Also presented at 1977 GSA Annual Meetings, Seattle, Wash.)
91. Walker, *Evolution of the Atmosphere*, p. 262.
92. H.D. Holland, 1975. Comment at a conference on The Early History of the Earth

(University of Leicester, England). In Walker, *Evolution of the Atmosphere*, p. 262.

93. M. Muir, 1975. Comment at conference on the Early History of Earth (University of Leicester, England). In Walker, *Evolution of the Atmosphere*, p. 262.
94. B.J. Skinner, Oct. 13-15, 1975. Invited oral summation of contributions at U.S. Geol. Survey Quartz-Pebble workshop, Golden, Colo.
95. D.E. Grandstaff, 1980. *Precambrian Res.* **13**, 1.
96. Ibid., p. 1.
97. E. Dimroth and M.M. Kimberley, 1976. *Can. J. Earth Sci.* **13**, 1161.

CHAPTER 6

Plausibility and Investigator Interference

Destruction of essential chemicals dominated our discussion of the prebiotic soup in Chapter 4. Re-examination of the early earth and its atmosphere in Chapter 5 shows it would have been far less reducing in character, and less conducive to abiogenic synthesis than previously imagined. If the theory of abiogenesis is to have any support, then *the burden to demonstrate such support rests squarely with the prebiotic simulation experiments*. And seemingly, reported results from simulation experiments suggest that a wide variety of important precursor chemicals would have existed in substantial concentrations in primitive water basins. Yet this contrasts sharply with the view presented in Chapter 4. Why the discrepancy? The answer becomes clear upon examining the details of prebiotic simulation experiments.

We propose in this chapter to evaluate various kinds of prebiotic simulation experiments (Chapter 3) and their associated techniques. Each of these techniques will be briefly discussed and some assessment of their geochemical plausibility offered. We provide this to point out the need for a criterion for the acceptable role of the investigator in prebiotic simulation experiments. We will then arrange these experimental techniques on a scale of increasing geochemical implausibility. This ordering necessarily involves questions of judgment and may be revised as time goes on.

Evaluation of Various Types of Simulation Experiments and Techniques

Simulation Experiments Using Ultraviolet Light

The successful synthesis of amino acids and other organic compounds using ultraviolet light has been reported in laboratory simulation experiments. These experiments used short-wavelength (i.e., $<$ 2000 Å) ultraviolet light but excluded the long-wavelength (i.e., $>$ 2000 Å)[1] UV which is so effective in destruction.[2] Although this practice is effective, it is dubious as a prebiotic simulation procedure, since the full solar spectrum would have irradiated the primitive earth.

Photosensitization

As we discussed in Chapter 3, photosensitization provides a means of using the plentiful longer-wavelength ultraviolet light (2000-3000 Å) to bring about photochemical reaction of the "primitive" reducing atmospheric gases. Mercury vapor, formaldehyde, and hydrogen sulfide gas all have served as photosensitizing agents, absorbing energy and transferring it to these primitive gases, thus enabling reactions to take place in the longer spectral region.[3]

A photosensitizer with an appropriate absorption spectrum can provide further benefits, too. For example, hydrogen sulfide can provide a protective shield against long-wavelength photodestruction of amino acids, as well as other biomonomers and essential intermediates produced in the atmosphere.[4] This protective shield operates because light in the range 2000-2600 Å is absorbed by hydrogen sulfide when it is present in sufficient concentration. Vulnerable organic molecules which otherwise would absorb below 2600 Å are thus protected.[5] Such a process operating in the primitive reducing atmosphere would have promoted the production and accumulation of vital precursors.

It is doubtful, however, that formaldehyde or hydrogen sulfide could have reached levels of concentration required to serve as early earth photosensitizers or to protect organic products from photodecomposition. For as it turns out, formaldehyde and hydrogen sulfide are themselves vulnerable to photodestruction, as previously mentioned, and no suitable shield appears to exist for them.

Of the two, hydrogen sulfide would be the most attractive candidate to serve the dual role of photosensitizer and shield. It would, however, have been photolyzed to free sulfur and hydrogen in only 10,000 years,[6] and there is no sufficient mechanism known for replenishing hydrogen sulfide.

The search for a suitable photosensitizer continues, but the field of candidates is limited. It must be assumed that such an agent was one of the simple gaseous components of the primitive atmosphere, or a derivative from it. Thus mercury vapor could not possibly have served generally as a photosensitizer on the early earth, although it might have had some localized application for short periods, as an effluent gas of volcanoes.[7] Photosensitization itself is not called into question, for photosynthesis uses chlorophyll as a photosensitizer enabling plants to utilize sunlight. But the use of this technique as a simulation precedure depends on geochemically implausible conditions. The pivotal question concerns whether system conditions necessary for photosensitization and shielding could reasonably obtain on an early earth.

Other Energy Sources: Heat

Experiments using heat, electrical discharge, and shock waves are also subject to criticism. Serious questions must be raised about the geological relevance of the heat experiments. For example, we do not find local high-temperature (> 150°C) regions on earth except for geologically brief periods of time. Volcanoes, fumaroles, steam spouts, etc. have been cited as heat energy sources, but they are generally too far apart geographically, and do not last over geologically significant times.[8] Scientists who accept heat as a legitimate source have usually argued that protocells at least originated very quickly and so brief geologic periods of energy inputs are all that are required. A continuous supply of intermediate chemicals was needed, however, until photosynthesis developed.[9] For this reason, it is believed by most scientists that only general sources of energy (e.g., ultraviolet light) could have been effective for the origin of life.

It has also been suggested that wind blowing the primitive gases over hot lava (500-1000°C) would subject them to high temperatures for brief periods. In the unconfined, natural situation, however, slightly warmed gases would rise quickly away from the hot lava, and thus never approach the temperature needed for reaction.[10] In more confined settings, such as pipes or fissures in rocks, the objection is that any organic molecules formed there would remain in the heat, and such sustained heating of organic materials would destroy them.[11]

Lightning

Electrical discharge experiments have attempted to simulate lightning on the early earth. The actual lightning leader is much too hot (i.e., 20,000°K) for effective synthesis, however, immediately destroying any products.[12] Much milder electrical discharges, the so-called corona discharges from pointed objects, have also been simulated in experiments. The energy density used in these experiments is, however, nine orders of magnitude too great to be called a simulation of natural phenomena.[13] In more imaginable terms the Miller spark experiment adds so much energy that "two days of sparking represent an energy input into the system comparable to some 40 million years on the surface of the primitive Earth."[14] Another geologically implausible feature of electrical discharge experiments is the fact that they are closed systems containing as much as 75% hydrogen.[15] (While they are begun with more plausible hydrogen concentrations, hydrogen is generated in the reaction and not allowed to escape as it would from an open system.)

Traps

All prebiotic heat,[16] electrical discharge,[17] and ultraviolet light[18] (including photosensitization) experiments use traps. Traps allow for greater yields of product from equilibrium reactions in which dissolution would otherwise far outweigh synthesis (i.e., $K_{eq} \ll 1$).[19] Traps function by continually removing the small fraction of product formed by the reactions. As products are removed from the zone of their formation, additional reaction is continuously required to reestablish equilibrium. In this way, reactions can be productively prolonged until one of the reactants is finally consumed.

This technique functions in accordance with Le Chatelier's Principle, which states that when a stress is applied to a chemical reaction at equilibrium, in this case by the trap, the reaction will shift in the direction that relieves the stress and reestablishes equilibrium. Like the practice of concentrating chemical reactants, this technique is a legitimate means of collapsing time to manageable amounts.

This removal process also shields the products from subsequent destruction by the energy source which produced them. However,

Carl Sagan has aptly commented on this shielding effect in the experiments:

> The problem we're discussing is a very general one. We use energy sources to make organic molecules. It is found that the same energy sources can destroy these organic molecules. The organic chemist has an understandable preference for removing the reaction products from the energy source before they are destroyed. But when we talk of the origin of life, I think we should not neglect the fact that degradation occurs as well as synthesis, and that the course of reaction may be different if the products are not preferentially removed. In reconstructing the origin of life, we have to *imagine reasonable scenarios which somehow avoid this difficulty*. (Emphasis added.)[20]

But even a brief scanning of published papers and symposium addresses on the topic demonstrates that there is no unanimity concerning such "reasonable scenarios." Instead, rebuttals and rejoinders to proposed solutions abound. Without reviewing the particulars of this dispute we simply note that it has been suggested that traps simulate a natural mechanism whereby rain washed these vital precursor heat, shock-wave, photo- and electro-products down to the ocean, where they were protected from the destructive rays of solar ultraviolet. How were these chemicals transported safely to the sea? It has been hypothesized that hydrogen sulfide gas, formaldehyde, mercury vapor, or some other photosensitizer was present in sufficient quantities in the primitive atmosphere (despite criticisms of photosensitizers discussed earlier) to allow substantial long wavelength ultraviolet synthesis. Since long-wavelength UV could penetrate to great atmospheric depths, this shifts the zone of synthesis for amino acids and other vulnerable organic molecules closer to the ocean surface. From there they would not have had far to flee to the ocean's protection.[21] Heat, electrical discharge, and shock wave syntheses would also have been operative at lower altitudes. Thus transport time would already be short for organic compounds produced by these sources. If appropriate photosensitizers were present to intercept the destructive ultraviolet, as the hypothesis suggests, organic compounds synthesized in the atmosphere would be further protected, giving them an even greater chance for survival.

In spite of these factors it is not at all clear that the ocean would have provided the shielding function of a trap. Laboratory traps are not usually exposed to long-wavelength ultraviolet light, which would be the case for the ocean, where UV light would penetrate some tens of meters beneath the surface.[22] Furthermore, ocean currents periodically surface even the deep water, thus exposing its organic contents to destructive ultraviolet light. Because of this it would seem that the ocean would have had much less in common with a trap than is usually suggested.

The Concerto Effect

Laboratory simulation experiments are usually carried out by employing one of various energy sources in isolation. This is a legitimate procedure since what is sought is the relative effect of each energy source. It is true, too, that the total effect is merely the sum of the effects of isolated energy sources. What often gets ignored, however, is that not only are the synthetic effects summed, but the destructive effects also. As we saw in Chapter 4, these energy sources act together or *in concert* in the natural situation, both in synthesis and destruction of organic compounds. One energy source destroys what another source produces. Destruction predominates!

Protection from energy sources is not the only concern. Many laboratory experiments use carefully selected, highly purified, and often concentrated reactants in solutions isolated from other constituents of the soup mixture. The practice of using concentrated chemicals is based on the well-known "law of mass action," which simply states that the rate of a chemical reaction is proportional to the concentration of the reacting substances. In other words, if a chemical reaction occurs slowly in dilute solution (viz., the primitive ocean), it will occur much more rapidly in concentrated solution (viz., the investigator's flask). In this way, investigators seek to compress into manageable laboratory time chemical reactions that normally would have taken millions of years.[23] The reactions are not thereby altered, but only hastened. There is merit to this practice then, even if natural concentrating mechanisms were not effective on the early earth. Many other features of laboratory simulation techniques, however, are suspect when viewed against the backdrop of Chapter 4.

Isolated Reactants

Practically all simulated ocean experiments reported in the scientific literature have been based on the assumption that if two or three chemicals react when isolated from the soup mixture, they will also react in the same way in the presence of diverse chemicals in the soup.

This assumption is seen in part of a discussion that took place in the *Proceedings of the First Conference on Origins of Life*, held in 1967.[24] Alex Rich asked Leslie Orgel whether he or others had "tried what I have called *Syntheses in the Whole*: that is to say, you have a spark discharge, a handful of sand, and lots of miscellaneous debris, and then you look for the production of cytosine, uracil, and so on."[25] Orgel responded: "This is the opposite of what we are trying to do. We believe you should learn the kinetics of each step, and when you think you understand it adequately, then try to put the thing together. We have not really gone to this later stage yet. We can get as far as purines quite easily. Sooner or later someone should do a giant experiment to try to do all the syntheses simultaneously, but I think it would be foolish to start that way."[26]

As we saw in Chapter 3, it is part of a general operating procedure to perform lab experiments which give some fair chance of disentangling the many individual reactions that would occur in the soup, to provide a reasonable way to discover reaction mechanisms and pathways.

In spite of the fact that the procedure of isolating reactants is almost universally used and assumed to be valid, for all practical purposes, this assumption is false in the general case. It is false because it overlooks the synergism of multiple reactions, the Concerto Effect. A mixture has a characteristic behavior of its own; it is not the simple sum of its individual components.[27] All components in a mixture have definite affinities for reacting with each other. Consequently, *soup mixture reactions do not equal the sum of the individual isolated reactions*. This has been seen in a great deal of the discussion in Chapter 4 about destructive interactions in the soup, and the scavenging mechanisms that "sweep clean" water basins of essential organic compounds. To state the case in general terms, substance A might react with substance B when isolated from substances C, D, and E. When all these substances are mixed together, however, competing reactions can be envisioned which assure that virtually no product accumulates from the reaction between A and B. Also, the reaction between A and B may begin as it would in isolation, only to be interrupted at some later step. Simulation experiments have thus produced some products which conceivably *would never occur* in the primitive soup.

To illustrate, consider whether freon (e.g., dichlorodifluoromethane) ever existed on this planet before a chemist synthesized it in a laboratory earlier in this century. It was of course possible, and a few molecules conceivably formed sometime in terrestrial history. In the *practical* sense, however, freon owes its existence to investigator intervention—the careful guidance of reactions down a specified chemical pathway.

Furthermore, on a primitive earth many chemicals would have been present that are usually absent in primitive atmosphere experiments. For example, aldehydes including reducing sugars would have been present, but these are not identified as products in primitive atmosphere simulation experiments. As a result, destructive interactions with amino acids are obviated and amino acids accumulate.*

This use of selected chemicals in simulation experiments is highly artificial, and creates a certain unrealism in our expectations of the early earth. In other words, when considering whether the ocean could have served as a trap, we must take into account the Concerto Effect, according to which the interaction of matter and energy must be considered synergistically.

Developing a Scale of Geochemical Plausibility

On the basis of the discussion here and in Chapter 4, we infer that the various simulation experiments can be ranked according to their geochemical plausibility (see fig. 6-1). We begin with the experimental reaction system of dilute solutions mixed together for a "synthesis in the whole" where the Concerto Effect is operative. This should form the basement of the scale, indicating the greatest geochemical plausibility of the various experiments examined. Next, the use of more concentrated solutions where the law of mass action would apply by extrapolation is only slightly less plausible than "Synthesis in the Whole."

Since it is conceivable that some as yet undiscovered mechanism worked to maintain hydrogen sulfide concentrations in the atmosphere, and since that alone would render photosensitization plausible, we place photosensitization next on the scale. It is certainly more plausible than using traps, for example, which would have required several gratuitous factors working simultaneously on the

*If amino acids were formed in spark discharge experiments by the Strecker synthesis (Chapter 3), then aldehydes would have been present. Aldehydes would have been consumed, however, through reactions with excess HCN. This interpretation is consistent with the fact that the major product in these experiments is formic acid, probably through the hydrolysis of HCN. The end result is that in spark discharge experiments, amino acids can accumulate in the trap precisely because there are no aldehydes left to react with them.

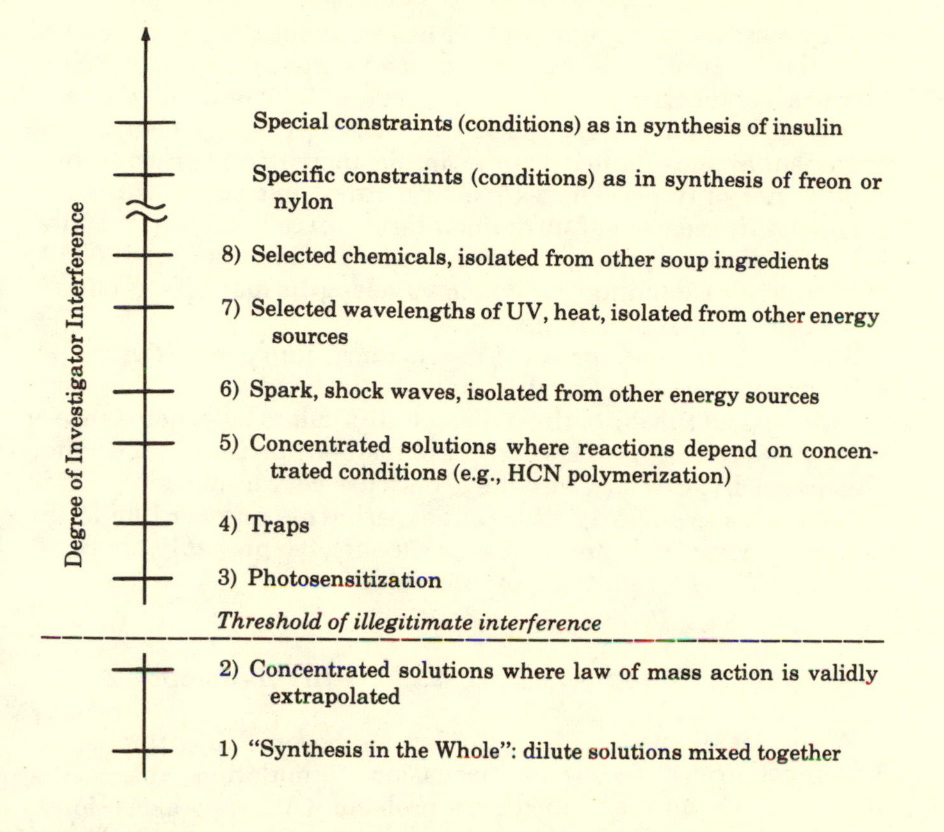

Figure 6-1.
Geochemical plausibility scale for evaluating prebiotic simulation experiments. Experimental techniques (conditions) are arranged according to the degree of investigator interference. At some point along the scale investigator involvement reaches a threshold, beyond which investigator interference is illegitimate.

early earth. More implausible still are those experiments which depend on conditions of higher concentration of reacting substances, e.g., HCN polymerization experiments, since there is greater question as to the existence of natural concentrating mechanisms.

Continuing up the scale, we come to spark and shock wave experiments, each used in isolation from other energy sources. We rank these experiments more implausible than those whose success is dependent on higher concentration of chemicals, because no conceivable natural means for isolating energy sources is known. Use of both heat and selected wavelengths of UV light is more implausible still. Not only is there the lack of means for isolating them from other energy sources, but greater doubt arises about their geochemical plausibility. It may be argued that using energy in spark experiments several orders of magnitude greater than could have existed on the early earth merely "speeds up" the process. No comparable argument applies for heat. For example, increasing temperature to 1000°C not only accelerates reaction rates, but destroys organic products. In the case of ultraviolet light, there is no natural filter known that would justify use of selected wavelengths (i.e., < 2000 Å) of light while excluding the longer wavelengths more destructive to some essential organic compounds.

Finally, to indicate greatest geochemical implausibility, we put experiments using selected chemicals, isolated from other soup ingredients, at the top of the scale. It is difficult to tell whether use of selected wavelengths of UV is more plausible than the use of isolated chemicals. In any case, we believe both are very implausible conditions. It does seem fairly clear that experiments number 1 and 2 are definitely acceptable prebiotic experiments, 3-6 probably unacceptable, and 7 and 8 definitely unacceptable.

Determining Acceptable Investigator Involvement

When does experimenter interference become illegitimate? As basic as this question is to the discussion of simulation experiments, it is very seldom mentioned as a problem. (A happy exception is Orgel and Lohrmann.[28]) Even when it is recognized, as with the use of high temperature and exotic chemicals, the discussion proceeds without any agreed-on criterion about what constitutes a legitimate simulation experiment. As a result, the discussion is surrounded by controversy. Throughout Chapter 4 we saw data showing that a wide discrepancy exists between plausible geochemical conditions and the conditions used in prebiotic simulation experiments. It is too radical to suggest that such experiments are without value. Their true value is difficult to assess, however.

Since all experiments are performed by an experimenter, they must involve *investigator intervention*. Yet experiments must be disqualified as prebiotic simulations when a certain class of investigator influence is *crucial* to their success. This is seen by analogy to the generally held requirement that no outside or supernatural agency was allowed to enter nature at the time of life's origin, was *crucial* to it, and then withdrew from history.[29] We can apply this principle through a careful extension of the analogy. In the preparation of a prebiotic simulation experiment, the investigator creates the setting, supplies the aqueous medium, the energy, the chemicals, and establishes the boundary conditions. This activity produces the general background conditions for the experiment, and while it is crucial to the success of the experiment, it is *quite legitimate* because it simulates plausible early earth conditions. The interference of the investigator becomes *crucial in an illegitimate sense*, however, wherever laboratory conditions are not warranted by analogy to reliably plausible features of the early earth itself.

Thus the illegitimate intervention of the investigator is directly proportional to the geochemical implausibility of the condition arising from experimental design and/or the investigator's procedure, the illegitimate interference being greatest when such plausibility is missing altogether.

With this in mind, it seems reasonable to suggest that permissible interference by the investigator would include developing plausible design features of the experiment, adjusting the initial reaction mixture, beginning the input of free energy to drive the reaction at the outset, and performing whatever minimal disturbance to the system is necessary to withdraw portions of the reaction products at various stages for analysis.

Usually, in laboratory experiments, an experimenter employs a host of manipulative interventions in an effort to guide natural processes down specific nonrandom chemical pathways. In other words it is the character of the constraint that determines the result. In some chemical syntheses, for example, it may be necessary to combine reactants in a particular order, or vary the rates of addition in order to control temperature, to adjust pH at a crucial color change, to remove products of reaction after ten minutes instead of twenty minutes, etc., etc. Such manipulations are the hallmark of intelligent, exogenous interference and *should not be employed in any prebiotic experiment.*

The arrangement of experimental techniques (conditions) in fig. 6-1 represents a scale or continuum of investigator interference. At some point on the scale, a degree of implausibility is reached where the experiment can no longer be considered acceptable. Beyond that point, there is no analogy between the techniques and reliably plausible prebiotic conditions. The experimenter who deviates from plausible conditions is like an actor who has forgotten his lines and begins to *ad-lib*. Such techniques constitute illegitimate interference, and cannot be given the same status as those lying within the threshold of acceptability.

In light of our study, we draw the line of legitimate interference between 2 and 3, i.e., between experiments using concentrated chemicals where the law of mass action is validly extrapolated and experiments using photosensitization. Both the relative ordering and the drawing of the line of acceptable interference are tentative. The principal purpose in presenting this scale, however, is to emphasize how important it is that a criterion for experiment acceptability be established.

Summary and Conclusion

Summarizing the above discussion it is our view that for each of the experimental techniques (conditions) listed as being above the line of crucial but acceptable interference, the investigator has played a highly significant but illegitimate role in experimental success. Brooks and Shaw have commented on this after a review of abiotic experiments:

> These experiments...claim abiotic synthesis for what has in fact been produced and designed by highly intelligent and very much biotic man.[30]

In other words, for each of the unacceptable experimental techniques, the investigator has established experimental constraints, imposing intelligent influence upon a supposedly "prebiotic earth." Where this informative intervention of the investigator is ignored, the illusion of prebiotic simulation is fostered. This unfortunate state of affairs will continue until the community of origin-of-life researchers agree on criteria for experiment acceptability.

If the techniques representing investigator interference are to be afforded the status of valid simulation, the burden must remain with the investigators to demonstrate their plausibility. This is nothing more than the demand of good science.

References

1. W.E. Groth and H. Weyssenhoff, 1957. *Naturwiss.* **44**, p. 510; 1959. *Ann. Physik* **4**, p. 69; 1960. *Planet. Space Sci.* **2**, p. 79; C. Ponnamperuma and J. Flores, 1966. *Abstr. Amer. Chem. Soc.*, 152nd Meeting Sept. 11-16, New York; A.N. Terenin, 1959. In *The Origin of Life on the Earth*, ed., A.I. Oparin, Oxford: Pergamon Press, p. 136; N. Dodonova and A.I. Sidorova, 1961. *Biophysics* **6**, 14.
2. H.R. Hulett, 1969. *J. Theoret. Biol.* **24**, 56; J.W.S. Pringle, 1954. *New Biology* No. 16, p. 54; C. Sagan, 1961. *Rad. Res.* **15**, 174; S. Miller, H. Urey, and J. Oro, 1976. *J. Mol. Evol.* **9**, 59.
3. C. Sagan and B.N. Khare, 1971. *Science* **173**, 417; Khare and Sagan, 1973. In *Molecules in the Galactic Environment*, eds. M.A. Gordon, and L.E. Snyder, New York: John Wiley, p. 399. K. Hong, J. Hong, and R. Becker, 1974. *Science* **184**, 984.
4. N. Friedmann, W.J. Haverland and S.L. Miller, 1971. In *Chemical Evolution and the Origin of Life*, eds. R. Buvet and C. Ponnamperuma. Amsterdam: North-Holland, p. 123.
5. Ibid., p. 129.
6. J.P. Ferris and D.E. Nicodem, 1974. In *The Origin of Life and Evolutionary Biochemistry*, eds. K. Dose, S.W. Fox, G. A. Deborin, and T.E. Pavlovskaya. New York: Plenum Press, p. 113.
7. S.M. Siegel and B.Z. Siegel, 1976. *Origins of Life* **2**, 175.
8. Stanley L. Miller and Harold C. Urey, 1959. *Science* **130**, 245; S.L. Miller and L.E. Orgel, 1974. *The Origins of Life on the Earth*. Englewood Cliffs, New Jersey: Prentice-Hall, p. 145; S.L. Miller, H.C. Urey, and J. Oro, 1976. *J. Mol. Evol.* **9**, 59.
9. Hulett, *J. Theoret. Biol.*, p. 56.
10. H.R. Hulett, 1973. In *Proceedings of the Fourth Conference on Origins of Life: Chemistry and Radioastronomy*, ed. Lynn Margulis. New York: Springer-Verlag, p. 95.
11. C. Sagan, in *Proceedings of the Fourth Conference on Origins of Life*, p. 95.
12. A. Bar-Nun, N. Bar-Nun, S.H. Bauer, and C. Sagan, 1970. *Science* **168**, 470.
13. Hulett, *J. Theoret. Biol.*, p. 64.
14. C.E. Folsome, 1979. *The Origin of Life*. San Francisco: W. H. Freeman & Co., p. 62.
15. S.W. Fox and K. Dose, 1977. *Molelcular Evolution and the Origin of Life*. New York: Marcel Dekker, p. 74-75.
16. K. Harada and S.W. Fox, 1965. In *The Origins of Prebiological Systems and of Their Molecular Matrices*, New York: Academic Press, p. 187.
17. S.L. Miller, 1955. *J. Am. Chem. Soc.* **77**, 2351.
18. Groth and Weyssenhoff, *Naturwiss.*, p. 510; *Ann. Physik*, p. 69; *Planet. Space Sci.*, p. 79; Ponnamperuma and Flores, *Abstr. Amer. Chem. Soc.*, 152nd Meeting; Terenin, in *The Origin of Life on the Earth*, p. 136; N.Y. Dodonova and A.I. Sidorova, *Biophysics* **6**, 14; Sagan and Khare, *Science*, p. 417; Khare and Sagan, in *Molecules in the Galactic Environment*, p. 399; Hong, Hong, and Becker, *Science*, p. 984.
19. H. Sisler, C. VanderWerf, and A. Davidson, 1967. *College Chemistry*, 3rd ed. New York: Macmillan, p. 326-7.
20. C. Sagan, in *The Origins of Prebiological Systems*, p. 195-6, in a discussion.
21. Sagan and Khare, *Science*, p. 417; Sagan, in *Proceedings of the Fourth Conference on Origins of Life*, p. 97.

22. L.V. Berkner and L.C. Marshall, 1965. *J. Atmos. Sci.* **22**, 225.
23. Dean H. Kenyon and Gary Steinman, 1969. *Biochemical Predestination*. New York: McGraw-Hill, p. 284.
24. Lynn Margulis, ed., 1970. *Origins of Life: Proceedings of the First Conference*, (Princeton, New Jersey, May 21-24, 1967.) New York: Gordon and Breach.
25. Alexander Rich, in *Origins of Life: Proceedings of the First Conference*, p. 183.
26. Leslie Orgel, in *Origins of Life: Proceedings of the First Conference*, p. 183.
27. A.I. Oparin, 1953. *The Origin of Life*, 2nd ed. transl. by S. Morgulis. New York: Dover, p. 146-7.
28. L.E. Orgel and R. Lohrmann, 1974. *Accounts of Chem. Res.* **7**, 368.
29. Kenyon and Steinman, *Biochemical Predestination*, p. 30.
30. J. Brooks and G. Shaw, 1973. *Origin and Development of Living Systems*. New York: Academic Press, p. 212.

CHAPTER 7

Thermodynamics of Living Systems

It is widely held that in the physical sciences the laws of thermodynamics have had a unifying effect similar to that of the theory of evolution in the biological sciences. What is intriguing is that the predictions of one seem to contradict the predictions of the other. The second law of thermodynamics suggests a progression from order to disorder, from complexity to simplicity, in the physical universe. Yet biological evolution involves a hierarchical progression to increasingly complex forms of living systems, seemingly in contradiction to the second law of thermodynamics. Whether this discrepancy between the two theories is only apparent or real is the question to be considered in the next three chapters. The controversy which is evident in an article published in the *American Scientist*[1] along with the replies it provoked demonstrates the question is still a timely one.

The First Law of Thermodynamics

Thermodynamics is an exact science which deals with energy. Our world seethes with transformations of matter and energy. Be these mechanical or chemical, the first law of thermodynamics—the principle of the Conservation of Energy—tells us that the total energy of the universe or any isolated part of it will be the same after any such transformation as it was before. A major part of the science of thermodynamics is accounting—giving an account of the energy of a system that has undergone some sort of transformation. Thus, we derive from the first law of thermodynamics that the change in the energy of a system (ΔE) is equal to the work done on (or by) the system (ΔW) and the heat flow into (or out of) the system (ΔQ). Mechanical work and energy are interchangeable, i.e., energy may be converted into mechanical work as in a steam engine, or mechanical work can be converted into energy as in the heating of a cannon which occurs as its barrel is bored. In mathematical terms (where the terms are as previously defined):

$$\Delta E = \Delta Q + \Delta W \tag{7-1}$$

The Second Law of Thermodynamics

The second law of thermodynamics describes the flow of energy in nature in processes which are irreversible. The physical significance of the second law of thermodynamics is that the energy flow in such processes is always toward a more uniform distribution of the energy of the universe. Anyone who has had to pay utility bills for long has become aware that too much of the warm air in his or her home during winter escapes to the outside. This flow of energy from the house to the cold outside in winter, or the flow of energy from the hot outdoors into the air-conditioned home in the summer, is a process described by the second law of thermodynamics. The burning of gasoline, converting energy-"rich" compounds (hydrocarbons) into energy-"lean" compounds, carbon dioxide (CO_2) and water (H_2O), is a second illustration of this principle.

The concept of entropy (S) gives us a more quantitative way to describe the tendency for energy to flow in a particular direction. The entropy change for a system is defined mathematically as the flow of energy divided by the temperature, or

$$\Delta S \geqslant \frac{\Delta Q}{T} \tag{7-2}$$

where ΔS is the change in entropy, ΔQ is the heat flow into or out of a system, and T is the absolute temperature in degrees Kelvin (K).*

*For a reversible flow of energy such as occurs under equilibrium conditions, the equality sign applies. For irreversible energy flow, the inequality applies.

A Driving Force

If we consider heat flow from a warm house to the outdoors on a cold winter night, we may apply equation 7-2 as follows:

$$\Delta S_T = \Delta S_{house} + \Delta S_{outdoors} \geqslant -\frac{\Delta Q}{T_1} + \frac{\Delta Q}{T_2} \qquad (7\text{-}3)$$

where ΔS_T is the total entropy change associated with this irreversible heat flow, T_1 is the temperature inside the house, and T_2 is the temperature outdoors. The negative sign of the first term notes loss of heat from the house, while the positive sign on the second term recognizes heat gained by the outdoors. Since it is warmer in the house than outdoors ($T_1 > T_2$), the total entropy will increase ($\Delta S_T > 0$) as a result of this heat flow. If we turn off the heater in the house, it will gradually cool until the temperature approaches that of the outdoors, i.e., $T_1 = T_2$. When this occurs, the entropy change (ΔS) associated with heat flow (ΔQ) goes to zero. Since there is no further driving force for heat flow to the outdoors, it ceases; equilibrium conditions have been established.

As this simple example shows, energy flow occurs in a direction that causes the total energy to be more uniformly distributed. If we think about it, we can also see that the entropy increase associated with such energy flow is proportional to the driving force for such energy flow to occur. The second law of thermodynamics says that the entropy of the universe (or any isolated system therein) is increasing; i.e., the energy of the universe is becoming more uniformly distributed.

It is often noted that the second law indicates that nature tends to go from order to disorder, from complexity to simplicity. If the most random arrangement of energy is a uniform distribution, then the present arrangement of the energy in the universe is nonrandom, since some matter is very rich in chemical energy, some in thermal energy, etc., and other matter is very poor in these kinds of energy. In a similar way, the arrangements of mass in the universe tend to go from order to disorder due to the random motion on an atomic scale produced by thermal energy. The diffusional processes in the solid, liquid, or gaseous states are examples of increasing entropy due to random atomic movements. Thus, increasing entropy in a system corresponds to increasingly random arrangements of mass and/or energy.

Entropy and Probability

There is another way to view entropy. The entropy of a system is a measure of the probability of a given arrangement of mass and energy within it. A statistical thermodynamic approach can be used to further quantify the system entropy. High entropy corresponds to high probability. As a random arrangement is highly probable, it would also be characterized by a large entropy. On the other hand, a highly ordered arrangement, being less probable, would represent a lower entropy configuration. The second law would tell us then that events which increase the entropy of the system require a change from more order to less order, or from less-random states to more-random states. We will find this concept helpful in Chapter 9 when we analyze condensation reactions for DNA and protein.

Clausius[2], who formulated the second law of thermodynamics, summarizes the laws of thermodynamics in his famous concise statement: "The energy of the universe is constant; the entropy of the universe tends toward a maximum." The universe moves from its less probable current arrangement (low entropy) toward its most probable arrangement in which the energy of the universe will be more uniformly distributed.

Life and the Second Law of Thermodynamics

How does all of this relate to chemical evolution? Since the important macromolecules of living systems (DNA, protein, etc.) are more energy rich than their precursors (amino acids, heterocyclic bases, phosphates, and sugars), classical thermodynamics would predict that such macromolecules will not spontaneously form.

Roger Caillois has recently drawn this conclusion in saying, "Clausius and Darwin cannot both be right."[3] This prediction of classical thermodynamics has, however, merely set the stage for refined efforts to understand life's origin. Harold Morowitz[4] and others have suggested that the earth is not an isolated system, since it is open to energy flow from the sun. Nevertheless, one cannot simply dismiss the problem of the origin of organization and complexity in biological systems by a vague appeal to open-system, non-equilibrium thermodynamics. The mechanisms responsible for the emergence and maintenance of coherent (organized) states must be defined. To clarify the role of mass and energy flow through a system as a *possible* solution to this problem, we will look in turn at the thermodynamics of (1) an isolated system, (2) a closed system, and (3) an open system. We will then discuss the application of open-system thermodynamics to living systems. In Chapter 8 we will apply the thermodynamic concepts presented in this chapter to the prebiotic synthesis of DNA and protein. In Chapter 9 this theoretical analysis will be used to interpret the various prebiotic synthesis experiments for DNA and protein, suggesting a physical basis for the uniform lack of success in synthesizing these crucial components for living cells.

Isolated Systems

An isolated system is one in which neither mass nor energy flows in or out. To illustrate such a system, think of a perfectly insulated thermos bottle (no heat loss) filled initially with hot tea and ice cubes. The total energy in this isolated system remains constant but the distribution of the energy changes with time. The ice melts and the energy becomes more uniformly distributed in the system. The initial distribution of energy into hot regions (the tea) and cold regions (the ice) is an ordered, nonrandom arrangement of energy, one not likely to be maintained for very long. By our previous definition then, we may say that the entropy of the system is initially low but gradually increases with time. Furthermore, the second law of thermodynamics says the entropy of the system will continue to increase until it attains some maximum value, which corresponds to the most probable state for the system, usually called equilibrium.

In summary, isolated systems always maintain constant total energy while tending toward maximum entropy, or disorder. In mathematical terms

$$\frac{\Delta E}{\Delta t} = 0$$

$$\text{(isolated system)} \qquad (7\text{-}4)$$

$$\frac{\Delta S}{\Delta t} \geqslant 0$$

where ΔE and ΔS are the changes in the system energy and system entropy respectively, for a time interval Δt. Clearly the emergence of order of any kind in an isolated system is not possible. The second law of thermodynamics says that an isolated system always moves in the direction of maximum entropy and, therefore, disorder.

It should be noted that the process just described is irreversible in the sense that once the ice is melted, it will not reform in the thermos. As a matter of fact, natural decay and the general tendency toward greater disorder are so universal that the second law of thermodynamics has been appropriately dubbed "time's arrow."[5]

Closed Systems near Equilibrium

A closed system is one in which the exchange of energy with the outside world is permitted but the exchange of mass is not. Along the boundary between the closed system and the surroundings, the temperature may be different from the system temperature, allowing energy flow into or out of the system as it moves toward equilibrium. If the temperature along the boundary is variable (in position but not time), then energy will flow *through* the system, maintaining it some distance from equilibrium. We will discuss closed systems near equilibrium first, followed by a discussion of closed systems removed from equilibrium next.

If we combine the first and second laws as expressed in equations 7-1 and 7-2 and replace the mechanical work term W by $-P\,\Delta V$*, where P is pressure and ΔV is volume change, we obtain

$$\Delta S \geqslant \frac{\Delta E + P\Delta V}{T} \qquad (7\text{-}5)$$

Algebraic manipulation gives

$$\Delta E + P\Delta V - T\Delta S \leqslant 0 \quad \text{or} \quad \Delta G \leqslant 0 \qquad (7\text{-}6)$$

where $$\Delta G = \Delta E + P\,\Delta V - T\Delta S$$

The term on the left side of the inequality in equation 7-6 is called the change in the Gibbs free energy (ΔG). It may be thought of as a

*Volume expansion ($\Delta V > 0$) corresponds to the system doing work, and therefore losing energy. Volume contraction ($\Delta V < 0$) corresponds to work being done on the system.

thermodynamic potential which describes the tendency of a system to change—e.g., the tendency for phase changes, heat conduction, etc. to occur. If a reaction occurs spontaneously, it is because it brings a decrease in the Gibbs free energy ($\Delta G < 0$). This requirement is equivalent to the requirement that the entropy of the universe increase. Thus, like an increase in entropy, a decrease in Gibbs free energy simply means that a system and its surroundings are changing in such a way that the energy of the universe is becoming more uniformly distributed.

We may summarize then by noting that the second law of thermodynamics requires

$$\frac{\Delta G}{\Delta t} \leqslant 0 \qquad \text{(closed system)} \qquad (7\text{-}7)$$

where Δt indicates the time period during which the Gibbs free energy changed.

The approach to equilibrium is characterized by

$$\frac{\Delta G}{\Delta t} \longrightarrow 0 \qquad \text{(closed system)} \qquad (7\text{-}8)$$

The physical significance of equation 7-7 can be understood by rewriting equations 7-6 and 7-7 in the following form:

$$\frac{\Delta S}{\Delta t} - \left[\frac{1}{T}\left(\frac{\Delta E}{\Delta t} + \frac{P\Delta V}{\Delta t}\right)\right] \geqslant 0 \qquad (7\text{-}9)$$

or

$$\frac{\Delta S}{\Delta t} - \frac{1}{T}\,\frac{\Delta H}{\Delta t} \geqslant 0$$

and noting that the first term represents the entropy change due to processes going on within the system and the second term represents the entropy change due to exchange of mechanical and/or thermal energy with the surroundings. This simply guarantees that the sum of the entropy change in the system and the entropy change in the surroundings will be greater than zero; i.e., the entropy of the universe must increase. For the isolated system, $\Delta E + P\Delta V = 0$ and equation 7-9 reduces to equation 7-4.

A simple illustration of this principle is seen in phase changes such as water transforming into ice. As ice forms, energy (80 calories/gm) is liberated to the surrounding. The change in the entropy of the system as the amorphous water becomes crystalline ice is −0.293 entropy units (eu)/degree Kelvin (K). The entropy change is negative because the thermal and configuration* entropy (or disorder) of water is greater than that of ice, which is a highly ordered crystal. Thus, the thermodynamic conditions under which water will transform to ice are seen from equation 7-9 to be:

$$-0.293 - \left(\frac{-80}{T}\right) > 0 \qquad (7\text{-}10a)$$

or

$$T \leqslant 273K \qquad (7\text{-}10b)$$

For condition of $T < 273K$ energy is removed from water to produce ice, and the aggregate disordering of the surroundings is greater than the ordering of the water into ice crystals. This gives a net increase in the entropy of the universe, as predicted by the second law of thermodynamics.

It has often been argued by analogy to water crystallizing to ice that simple monomers may polymerize into complex molecules such as protein and DNA. The analogy is clearly inappropriate, however. The $\Delta E + P\Delta V$ term (equation 7-9) in the polymerization of important organic molecules is generally positive (5 to 8 kcal/mole), indicating the reaction can never spontaneously occur at or near equilibrium.† By contrast the $\Delta E + P\Delta V$ term in water changing to ice is a negative, −1.44 kcal/mole, indicating the phase change is spontaneous as long as $T < 273K$, as previously noted. The atomic bonding forces draw water molecules into an orderly crystalline array when the thermal agitation (or entropy driving force, $T \Delta S$) is made sufficiently small by lowering the temperature. Organic monomers such as amino acids resist combining at all at any temperature, however, much less in some orderly arrangement.

Morowitz[6] has estimated the increase in the chemical bonding energy as one forms the bacterium *Escherichia coli* from simple precursors to be 0.0095 erg, or an average of 0.27 ev/atom for the 2 x

---

*Configurational entropy measures randomness in the distribution of matter in much the same way that thermal entropy measures randomness in the distribution of energy.

†If $\Delta E + P\Delta V$ is positive, the entropy term in eq. 7-9 must be negative due to the negative sign which preceeds it. The inequality can only be satisfied by ΔS being sufficiently positive, which implies disordering.

10^{10} atoms in a single bacterial cell. This would be thermodynamically equivalent to having water in your bathtub spontaneously heat up to 360°C, happily a most unlikely event. He goes on to estimate the probability of the spontaneous formation of one such bacterium in the entire universe in five billion years *under equilibrium conditions* to be $10^{-10^{11}}$. Morowitz summarizes the significance of this result by saying that "if equilibrium processes alone were at work, the largest possible fluctuation in the history of the universe is likely to have been no longer than a small peptide."[7] Nobel Laureate I. Prigogine et al., have noted with reference to the same problem that:

> The probability that at ordinary temperatures a macroscopic number of molecules is assembled to give rise to the highly ordered structures and to the coordinated functions characterizing living organisms is vanishingly small. The idea of spontaneous genesis of life in its present form is therefore highly improbable, even on the scale of billions of years during which prebiotic evolution occurred.[8]

It seems safe to conclude that systems near equilibrium (whether isolated or closed) can *never* produce the degree of complexity intrinsic in living systems. Instead, they will move spontaneously toward maximizing entropy, or randomness. Even the postulate of long time periods does not solve the problem, as "time's arrow" (the second law of thermodynamics) points in the wrong direction; i.e., toward equilibrium. In this regard, H.F. Blum has observed:

> The second law of thermodynamics would have been a dominant directing factor in this case [of chemical evolution]; the reactions involved tending always toward equilibrium, that is, toward less free energy, and, in an inclusive sense, greater entropy. From this point of view the lavish amount of time available should only have provided opportunity for *movement in the direction of equilibrium*.[9] (Emphasis added.)

Thus, reversing "time's arrow" is what chemical evolution is all about, and this will not occur in isolated or closed systems near equilibrium.

The possibilities are potentially more promising, however, if one considers a system subjected to energy flow which may maintain it far from equilibrium, and its associated disorder. Such a system is said to be a *constrained* system, in contrast to a system at or near equilibrium which is unconstrained. The possibilities for ordering in such a system will be considered next.

Closed Systems Far from Equilibrium

Energy flow through a system is the equivalent to doing work continuously on the system to maintain it some distance from equilibrium. Nicolis and Prigogine[10] have suggested that the entropy change (ΔS) in a system for a time interval (Δt) may be divided into two components.

$$\Delta S = \Delta S_e + \Delta S_i \qquad (7\text{-}11)$$

where ΔS_e is the entropy flux due to energy flow through the system, and ΔS_i is the entropy production inside the system due to irreversible processes such as diffusion, heat conduction, heat production, and chemical reactions. We will note when we discuss open systems in the next section that ΔS_e includes the entropy flux due to mass flow through the system as well. The second law of thermodynamics requires

$$\Delta S_i \geqslant 0 \qquad (7\text{-}12)$$

In an isolated system, $\Delta S_e = 0$ and equations 7-11 and 7-12 give

$$\Delta S = \Delta S_i \geqslant 0 \qquad (7\text{-}13)$$

Unlike ΔS_i, ΔS_e in a closed system does not have a definite sign, but depends entirely on the boundary constraints imposed on the system. The total entropy change in the system can be negative (i.e., ordering within system) when

$$\Delta S_e \leqslant 0 \text{ and } |\Delta S_e| > \Delta S_i \qquad (7\text{-}14)$$

Under such conditions a state that would normally be highly improbable under equilibrium conditions can be maintained indefinitely. It would be highly unlikely (i.e., statistically just short of impossible) for a disconnected water heater to produce hot water. Yet when the gas is connected and the burner lit, the system is constrained by energy flow and hot water is produced and maintained indefinitely as long as energy flows through the system.

An open system offers an additional possibility for ordering—that of maintaining a system far from equilibrium via mass flow through the system, as will be discussed in the next section.

Open Systems

An open system is one which exchanges both energy and mass with the surroundings. It is well illustrated by the familiar internal combustion engine. Gasoline and oxygen are passed through the system, combusted, and then released as carbon dioxide and water. The energy released by this mass flow through the system is converted into useful work; namely, torque supplied to the wheels of the automobile. A coupling mechanism is necessary, however, to allow the released energy to be converted into a particular kind of work. In an analagous way the dissipative (or disordering) processes ($\triangle S_i$) within an open system can be offset by a steady supply of energy to provide for $\triangle S_e$ type work. Equation 7-11, applied earlier to closed systems far from equilibrium, may also be applied to open systems. In this case, the $\triangle S_e$ term represents the negative entropy, or organizing work done on the system as a result of both energy and mass flow through the system. This work done to the system can move it far from equilibrium, maintaining it there as long as the mass and/or energy flow are not interrupted. This is an essential characteristic of living systems as will be seen in what follows.

Thermodynamics of Living Systems

Living systems are composed of complex molecular configurations whose total bonding energy is less negative than that of their chemical precursors (e.g., Morowitz's estimate of $\triangle E = 0.27$ ev/atom) and whose thermal and configurational entropies are also less than that of their chemical precursors. Thus, the Gibbs free energy of living systems (see equation 7-6) is quite high relative to the simple compounds from which they are formed. The formation and maintenance of living systems at energy levels well removed from equilibrium requires continuous work to be done on the system, even as maintenance of hot water in a water heater requires that continuous work be done on the system. Securing this continuous work requires energy and/or mass flow through the system, apart from which the system will return to an equilibrium condition (lowest Gibbs free energy, see equations 7-7 and 7-8) with the decomposition of complex molecules into simple ones, just as the hot water in our water heater returns to room temperature once the gas is shut off.

In living plants, the energy flow through the system is supplied principally by solar radiation. In fact, leaves provide relatively large surface areas per unit volume for most plants, allowing them to "capture" the necessary solar energy to maintain themselves far from equilibrium. This solar energy is converted into the necessary useful work (negative ΔS_e in equation 7-11) to maintain the plant in its complex, high-energy configuration by a complicated process called photosynthesis. Mass, such as water and carbon dioxide, also flows through plants, providing necessary raw materials, but not energy. In collecting and storing useful energy, plants serve the entire biological world.

For animals, energy flow through the system is provided by eating high energy biomass, either plant or animal. The breaking down of this energy-rich biomass, and the subsequent oxidation of part of it (e.g., carbohydrates), provides a continuous source of energy as well as raw materials. If plants are deprived of sunlight or animals of food, dissipation within the system will surely bring death. Maintenance of the complex, high-energy condition associated with life is not possible apart from a continuous source of energy. A source of energy alone is not sufficient, however, to explain the origin or maintenance of living systems. The additional crucial factor is *a means of converting this energy* into the necessary useful work to build and maintain complex living systems from the simple biomonomers that constitute their molecular building blocks.

An automobile with an internal combustion engine, transmission, and drive chain provides the necessary mechanism for converting the energy in gasoline into comfortable transportation. Without such an "energy converter," however, obtaining transportation from gasoline would be impossible. In a similar way, food would do little for a man whose stomach, intestines, liver, or pancreas were removed. Without these, he would surely die even though he continued to eat. Apart from a mechanism to couple the available energy to the necessary work, high-energy biomass is insufficient to sustain a living system far from equilibrium. In the case of living systems such a coupling mechanism channels the energy along specific chemical pathways to accomplish a very specific type of work. We therefore conclude that, given the availability of energy *and* an appropriate coupling mechanism, the maintenance of a living system far from equilibrium presents no thermodynamic problems.

In mathematical formalism, these concepts may be summarized as follows:

(1) The second law of thermodynamics requires only that the

entropy production due to irreversible processes within the system be greater than zero; i.e.,

$$\Delta S_i > 0 \quad (7\text{-}15)$$

(2) The maintenance of living systems requires that the energy flow through the system be of sufficient magnitude that the negative entropy production rate (i.e., useful work rate) that results be greater than the rate of dissipation that results from irreversible processes going on within the systems; i.e.,

$$|\Delta S_e| > \Delta S_i \quad (7\text{-}16)$$

(3) The negative entropy generation must be coupled into the system in such a way that the resultant work done is directed toward restoration of the system from the disintegration that occurs naturally and is described by the second law of thermodynamics; i.e.,

$$-\Delta S_e = \Delta S_i \quad (7\text{-}17)$$

where ΔS_e and ΔS_i refer not only to the magnitude of entropy change but also to the specific changes that occur in the system associated with this change in entropy. The coupling must produce not just any kind of ordering but the specific kind required by the system.

While the maintenance of living systems is easily rationalized in terms of thermodynamics, the *origin* of such living systems is quite another matter. Though the earth is open to energy flow from the sun, the means of converting this energy into the necessary work to build up living systems from simple precursors remains at present unspecified (see equation 7-17). The "evolution" from biomonomers to fully functioning cells is the issue. Can one make the incredible jump in energy and organization from raw material and raw energy, apart from some means of directing the energy flow through the system? In Chapters 8 and 9 we will consider this question, limiting our discussion to two small but crucial steps in the proposed evolutionary scheme namely, the formation of protein and DNA from their precursors.

It is widely agreed that both protein and DNA are essential for living systems and indispensable components of every living cell today.[11] Yet they are only produced by living cells. Both types of molecules are much more energy and information rich than the biomonomers from which they form. Can one reasonably predict their occurrence given the necessary biomonomers and an energy source? Has this been verified experimentally? These questions will be considered in Chapters 8 and 9.

References

1. Victor F. Weisskopf, 1977. *Amer. Sci.* **65**, 405-11.
2. R. Clausius, 1855. *Ann. Phys.,* **125**, 353.
3. R. Caillois, 1976. *Coherences Aventureuses*. Paris: Gallimard.
4. H.J. Morowitz, 1968. *Energy Flow in Biology*. New York: Academic Press, p. 2-3.
5. H.F. Blum, 1951. *Time's Arrow and Evolution*. Princeton: Princeton University Press.
6. H.J. Morowitz, *Energy Flow*, p. 66.
7. H.J. Morowitz, *Energy Flow*, p. 68.
8. I. Prigogine, G. Nicolis, and A. Babloyantz, November, 1972. *Physics Today*, p. 23.
9. H.F. Blum, 1955. *American Scientist* **43**, 595.
10. G. Nicolis and I. Prigogine, 1977. *Self-Organization in Nonequilibrium Systems*. New York: John Wiley, p. 24.
11. S.L. Miller and L.E. Orgel, 1974. *The Origins of Life on the Earth*. Englewood Cliffs, New Jersey: Prentice-Hall, p. 162-3.

CHAPTER 8

Thermodynamics and the Origin of Life

Peter Molton has defined life as "regions of order which use energy to maintain their organization against the disruptive force of entropy."[1] In Chapter 7 it has been shown that energy and/or mass flow through a system can constrain it far from equilibrium, resulting in an increase in order. Thus, it is thermodynamically possible to develop complex living forms, assuming the energy flow through the system can somehow be effective in organizing the simple chemicals into the complex arrangements associated with life.

In existing living systems, the coupling of the energy flow to the organizing "work" occurs through the metabolic motor of DNA, enzymes, etc. This is analogous to an automobile converting the chemical energy in gasoline into mechanical torque on the wheels. We can give a thermodynamic account of how life's metabolic motor works. The origin of the metabolic motor (DNA, enzymes, etc.) itself, however, is more difficult to explain thermodynamically, since a mechanism of coupling the energy flow to the organizing work is unknown for prebiological systems. Nicolis and Prigogine summarize the problem in this way:

> Needless to say, these simple remarks cannot suffice to solve the problem of biological order. One would like not only to establish that the second law ($dS_i \geq 0$) is compatible with a decrease in overall entropy ($dS < 0$), but also to indicate the mechanisms responsible for the emergence and maintenance of coherent states.[2]

Without a doubt, the atoms and molecules which comprise living cells individually obey the laws of chemistry and physics, including the laws of thermodynamics. The enigma is the origin of so unlikely an organization of these atoms and molecules. The electronic computer provides a striking analogy to the living cell. Each component in a computer obeys the laws of electronics and mechanics. The key to the computer's marvel lies, however, in the highly unlikely organization of the parts which harness the laws of electronics and mechanics. In the computer, this organization was specially arranged by the designers and builders and continues to operate (with occasional frustrating lapses) through the periodic maintenance of service engineers.

Living systems have even greater organization. The problem then, that molecular biologists and theoretical physicists are addressing, is how the organization of living systems could have arisen spontaneously. Prigogine et al., have noted:

> All these features bring the scientist a wealth of new problems. In the first place, one has systems that have evolved spontaneously to extremely organized and complex forms. Coherent behavior is really the characteristic feature of biological systems.[3]

In this chapter we will consider only the problem of the *origin* of living systems. Specifically, we will discuss the arduous task of using simple biomonomers to construct complex polymers such as DNA and protein by means of thermal, electrical, chemical, or solar energy. We will first specify the nature and magnitude of the "work"* to be done in building DNA and enzymes. In Chapter 9 we will describe the various theoretical models which attempt to explain how the undirected flow of energy through simple chemicals can accomplish the work necessary to produce complex polymers. Then we will review the experimental studies that have been conducted to test these models. Finally we will summarize the current understanding of this subject.

How can we specify in a more precise way the work to be done by energy flow through the system to synthesize DNA and protein from simple biomonomers? While the origin of living systems involves

*"Work in physics normally refers to force times displacement. In this chapter it refers in a more general way to the change in Gibbs free energy of the system that accompanies the polymerization of monomers into polymers.

more than the genesis of enzymes and DNA, these components are essential to any system if replication is to occur. It is generally agreed that natural selection can act only on systems capable of replication. This being the case, the formation of a DNA/enzyme system by processes other than natural selection is a necessary (though not sufficient) part of a naturalistic explanation for the origin of life.*

Order vs. Complexity in the Question of Information

Only recently has it been appreciated that the distinguishing feature of living systems is complexity rather than order.[4] This distinction has come from the observation that the essential ingredients for a replicating system—enzymes and nucleic acids—are all information-bearing molecules. In contrast, consider crystals. They are very orderly, spatially periodic arrangements of atoms (or molecules) but they carry very little information. Nylon is another example of an orderly, periodic polymer (a polyamide) which carries little information. Nucleic acids and protein are *aperiodic* polymers, and this aperiodicity is what makes them able to carry much more information. By definition then, *a periodic structure has order. An aperiodic structure has complexity*. In terms of information, periodic polymers (like nylon) and crystals are analogous to a book in which the same sentence is repeated throughout. The arrangement of "letters" in the book is highly ordered, but the book contains little information since the information presented—the single word or sentence—is highly redundant.

It should be noted that aperiodic polypeptides or polynucleotides do not *necessarily* represent meaningful information or biologically useful functions. A random arrangement of letters in a book is aperiodic but contains little if any useful information since it is devoid of meaning.† Only certain sequences of letters correspond to sentences, and only certain sequences of sentences correspond to paragraphs, etc. In the same way only certain sequences of amino acids in polypeptides and bases along polynucleotide chains corres-

---

*A sufficient explanation for the origin of life would also require a model for the formation of other critical cellular components, including membranes, and their assembly.

†H.P. Yockey, personal communication, 9/29/82. Meaning is extraneous to the sequence, arbitrary, and depends on some symbol convention. For example, the word "gift," which in English means a *present* and in German *poison*, in French is meaningless.

pond to useful biological functions. Thus, informational macromolecules may be described as being aperiodic and in a *specified* sequence.[5] Orgel notes:

> Living organisms are distinguished by their specified complexity. Crystals such as granite fail to qualify as living because they lack complexity; mixtures of random polymers fail to qualify because they lack specificity.[6]

Three sets of letter arrangements show nicely the difference between order and complexity in relation to information:

1. An ordered (periodic) and therefore specified arrangement:

   THE END THE END THE END THE END*

   Example: Nylon, or a crystal.

2. A complex (aperiodic) unspecified arrangement:

   AGDCBFE GBCAFED ACEDFBG

   Example: Random polymers (polypeptides).

3. A complex (aperiodic) specified arrangement:

   THIS SEQUENCE OF LETTERS CONTAINS A MESSAGE!

   Example: DNA, protein.

Yockey[7] and Wickens[8] develop the same distinction, explaining that "order" is a statistical concept referring to regularity such as might characterize a series of digits in a number, or the ions of an inorganic crystal. On the other hand, "organization" refers to physical systems and the specific set of spatio-temporal and functional relationships among their parts. Yockey and Wickens note that informational macromolecules have a low degree of order but a high degree of specified complexity. In short, the redundant order of

*Here we use "THE END" even though there is no reason to suspect that nylon or a crystal would carry even this much information. Our point, of course, is that even if they did, the bit of information would be drowned in a sea of redundancy.

crystals cannot give rise to specified complexity of the kind or magnitude found in biological organization; attempts to relate the two have little future.

Information and Entropy

There is a general relationship between information and entropy. This is fortunate because it allows an analysis to be developed in the formalism of classical thermodynamics, giving us a powerful tool for calculating the work to be done by energy flow through the system to synthesize protein and DNA (if indeed energy flow is capable of producing information). The information content in a given sequence of units, be they digits in a number, letters in a sentence, or amino acids in a polypeptide or protein, depends on the minimum number of instructions needed to specify or describe the structure. Many instructions are needed to specify a *complex*, information-bearing structure such as DNA. Only a few instructions are needed to specify an *ordered* structure such as a crystal. In this case we have a description of the initial sequence or unit arrangement which is then repeated *ad infinitum* according to the packing instructions.

Orgel[9] illustrates the concept in the following way. To describe a crystal, one would need only to specify the substance to be used and the way in which the molecules were to be packed together. A couple of sentences would suffice, followed by the instructions "and keep on doing the same," since the packing sequence in a crystal is regular. The description would be about as brief as specifying a DNA-like polynucleotide with a random sequence. Here one would need only to specify the proportions of the four nucleotides in the final product, along with instructions to assemble them randomly. The chemist could then make the polymer with the proper composition but with a random sequence.

It would be quite impossible to produce a correspondingly simple set of instructions that would enable a chemist to synthesize the DNA of an *E. coli* bacterium. In this case *the sequence matters*. Only by specifying the sequence letter-by-letter (about 4,000,000 instructions) could we tell a chemist what to make. Our instructions would occupy not a few short sentences, but a large book instead!

Brillouin,[10] Schrodinger,[11] and others[12] have developed both qualitative and quantitative relationships between information and entropy. Brillouin[13] states that the entropy of a system is given by

$$S = k \ln \Omega \qquad (8\text{-}1)$$

where S is the entropy of the system, k is Boltzmann's constant, and Ω corresponds to the number of ways the energy and mass in a system may be arranged.

We will use S_{th} and S_c to refer to the thermal and configurational entropies, respectively. Thermal entropy, S_{th}, is associated with the distribution of energy in the system. Configurational entropy S_c is concerned only with the arrangement of mass in the system, and, for our purposes, we shall be especially interested in the sequencing of amino acids in polypeptides (or proteins) or of nucleotides in polynucleotides (e.g., DNA). The symbols Ω_{th} and Ω_c refer to the number of ways energy and mass, respectively, may be arranged in a system.

Thus we may be more precise by writing

$$S = k \ln \Omega_{th}\Omega_c = k \ln \Omega_{th} + k \ln \Omega_c = S_{th} + S_c \qquad (8\text{-}2a)$$

where

$$S_{th} = k \ln \Omega_{th} \qquad (8\text{-}2b)$$

and

$$S_c = k \ln \Omega_c \qquad (8\text{-}2c)$$

Determining Information: From a Random Polymer to an Informed Polymer

If we want to convert a random polymer into an informational molecule, we can determine the increase in information (as defined by Brillouin) by finding the difference between the negatives of the entropy states for the initial random polymer and the informational molecule:

$$I = -(S_{cm} - S_{cr}) \qquad (8\text{-}3a)$$

$$I = S_{cr} - S_{cm} \qquad (8\text{-}3b)$$

$$= k \ln \Omega_{cr} - k \ln \Omega_{cm} \qquad (8\text{-}3c)$$

In this equation, I is a measure of the information content of an aperiodic (complex) polymer with a specified sequence, S_{cm} represents the configurational "coding" entropy of this polymer informed with a given message, and S_{cr} represents the configurational entropy of the same polymer for an unspecified or random sequence.*

Note that the information in a sequence-specified polymer is maximized when the mass in the molecule could be arranged in many different ways, only one of which communicates the intended message. (There is a large S_{cr} from eq. 8-2c since Ω_{cr} is large, yet $S_{cm} = 0$ from eq. 8-2c since $\Omega_{cm} = 1$.) The information carried in a crystal is small because S_c is small (eq. 8-2c) for a crystal. There simply is very little potential for information in a crystal because its matter can be distributed in so few ways. The random polymer provides an even starker contrast. It bears *no* information because S_{cr}, although large, is equal to S_{cm} (see eq. 8-3b).

In summary, equations 8-2c and 8-3c quantify the notion that only specified, aperiodic macromolecules are capable of carrying the large amounts of information characteristic of living systems. Later we will calculate "Ω_c" for both random and specified polymers so that the configurational entropy change required to go from a random to a specified polymer can be determined. In the next section we will consider the various components of the total work required in the formation of macromolecules such as DNA and protein.

DNA and Protein Formation:

Defining the Work

There are three distinct components of work to be done in assembling simple biomonomers into a complex (or aperiodic) linear polymer with a specified sequence as we find in DNA or protein. The change in the Gibbs free energy, ΔG, of the system during polymerization defines the total work that must be accomplished by energy flow through the system. The change in Gibbs free energy has previously been shown to be

$$\Delta G = \Delta E + P\,\Delta V - T\,\Delta S \quad (8\text{-}4a)$$

or

$$\Delta G = \Delta H - T\,\Delta S \quad (8\text{-}4b)$$

*Yockey and of Wickens define information slightly differently than Brillouin, whose definition we use in our analysis. The difference is unimportant insofar as our analysis here is concerned.

where a decrease in Gibbs free energy for a given chemical reaction near equilibrium guarantees an increase in the entropy of the universe as demanded by the second law of thermodynamics.

Now consider the components of the Gibbs free energy (eq. 8-4b) where the change in enthalpy (ΔH) is principally the result of changes in the total bonding energy (ΔE), with the ($P \Delta V$) term assumed to be negligible. We will refer to this enthalpy component (ΔH) as the *chemical work*. A further distinction will be helpful. The change in the entropy (ΔS) that accompanies the polymerization reaction may be divided into two distinct components which correspond to the changes in the thermal energy distribution (ΔS_{th}) and the mass distribution (ΔS_c), eq. 8-2. So we can rewrite eq. 8-4b as follows:

$$\Delta G = \Delta H - T\Delta S_{th} - T\Delta S_c \qquad (8\text{-}5)$$

Gibbs free energy	Chemical work	Thermal entropy work	Configurational entropy work

It will be shown that polymerization of macromolecules results in a decrease in the thermal and configurational entropies ($\Delta S_{th} < 0$, $\Delta S_c < 0$). These terms effectively increase ΔG, and thus represent additional components of work to be done beyond the chemical work.

Consider the case of the formation of protein or DNA from biomonomers in a chemical soup. For computational purposes it may be thought of as requiring two steps: (1) polymerization to form a chain molecule with an aperiodic but near-random sequence,* and (2) rearrangement to an aperiodic, specified information-bearing sequence. The entropy change (ΔS) associated with the first step is essentially all thermal entropy change (ΔS_{th}), as discussed above. The entropy change of the second step is essentially all configurational entropy change (ΔS_c). In fact, as previously noted, the change in configurational entropy (ΔS_c) = ΔS_c "coding" as one goes from a random arrangement (S_{cr}) to a specified sequence (S_{cm}) in a macromolecule is numerically equal to the negative of the information content of the molecule as defined by Brillouin (see eq. 8-3a).

In summary, the formation of complex biological polymers such as DNA and protein involves changes in the chemical energy, ΔH, the thermal entropy, ΔS_{th}, and the configurational entropy, ΔS_c, of

---

*Some intersymbol influence arising from differential atomic bonding properties makes the distribution of matter not quite random. (H.P. Yockey, 1981. *J. Theoret. Biol.* **91**, 13.)

the system. Determining the magnitudes of these individual changes using experimental data and a few calculations will allow us to quantify the magnitude of the required work potentially to be done by energy flow through the system in synthesizing macromolecules such as DNA and protein.

Quantifying the Various Components of Work

1. Chemical Work

The polymerization of amino acids to polypeptides (protein) or of nucleotides to polynucleotides (DNA) occurs through condensation reactions. One may calculate the enthalpy change in the formation of a dipeptide from amino acids to be 5 - 8 kcal/mole for a variety of amino acids, using data compiled by Hutchens.[14] Thus, chemical work must be done on the system to get polymerization to occur. Morowitz[15] has estimated more generally that the chemical work, or average increase in enthalpy, for macromolecule formation in living systems is 16.4 cal/gm. Elsewhere in the same book he says that the average increase in bonding energy in going from simple compounds to an *E. coli* bacterium is 0.27 ev/atom. One can easily see that chemical work must be done on the biomonomers to bring about the formation of macromolecules like those that are essential to living systems. By contrast, amino acid formation from simple reducing atmosphere gases (methane, ammonia, water) has an associated enthalpy change ($\triangle H$) of −50 kcal/mole to −250 kcal/mole,[16] which means energy is released rather than consumed. This explains why amino acids form with relative ease in prebiotic simulation experiments. On the other hand, forming amino acids from less-reducing conditions (i.e., carbon dioxide, nitrogen, and water) is known to be far more difficult experimentally. This is because the enthalpy change ($\triangle H$) is positive, meaning energy is required to drive the energetically unfavorable chemical reaction forward.

2. Thermal Entropy Work

Wickens[17] has noted that polymerization reactions will reduce the number of ways the translational energy may be distributed, while generally increasing the possibilities for vibrational and rotational energy. A net decrease results in the number of ways the thermal energy may be distributed, giving a decrease in the thermal entropy according to eq. 8-2b (i.e., $\Delta S_{th} < 0$). Quantifying the magnitude of this decrease in thermal entropy (ΔS_{th}) associated with the formation of a polypeptide or a polynucleotide is best accomplished using experimental results.

Morowitz[18] has estimated that the average decrease in thermal entropy that occurs during the formation of macromolecules of living systems in 0.218 cal/deg-gm or 65 cal/gm at 298K. Recent work by Armstrong et al.,[19] for nucleotide oligomerization of up to a pentamer indicates ΔH and $-T\ \Delta S_{th}$ values of 11.8 kcal/mole and 15.6 kcal/mole respectively, at 294K. Thus the decrease in thermal entropy during the polymerization of the macromolecules of life increases the Gibbs free energy and the work required to make these molecules, i.e., $-T\ \Delta S_{th} > 0$.

3. Configurational Entropy Work

Finally, we need to quantify the configurational entropy change (ΔS_c) that accompanies the formation of DNA and protein. Here we will not get much help from standard experiments in which the equilibrium constants are determined for a polymerization reaction at various temperatures. Such experiments do not consider whether a specific sequence is achieved in the resultant polymers, but only the concentrations of randomly sequenced polymers (i.e., polypeptides) formed. Consequently, they do not measure the configurational entropy (ΔS_c) contribution to the total entropy change (ΔS). However, the magnitude of the configurational entropy change associated with sequencing the polymers can be calculated.

Using the definition for configurational "coding" entropy given in eq. 8-2c, it is quite straightforward to calculate the configurational entropy change for a given polymer. The number of ways the mass of the linear system may be arranged (Ω_c) can be calculated using statistics. Brillouin[20] has shown that the number of distinct sequences one can make using N different symbols and Fermi-Dirac statistics is given by

$$\Omega = N! \qquad (8\text{-}6)$$

If some of these symbols are redundant (or identical), then the number of unique or distinguishable sequences that can be made is reduced to

$$\Omega_c = \frac{N!}{n_1!n_2!\ldots n_i!} \qquad (8\text{-}7)$$

where $n_1 + n_2 + ... + n_i = N$ and i defines the number of distinct symbols. For a protein, it is i = 20, since a subset of twenty distinctive types of amino acids is found in living things, while in DNA it is i = 4 for the subset of four distinctive nucleotides. A typical protein would have 100 to 300 amino acids in a specific sequence, or N = 100 to 300. For DNA of the bacterium *E. coli*, N = 4,000,000. In Appendix 1, alternative approaches to calculating Ω_c are considered and eq. 8-7 is shown to be a lower bound to the actual value.

For a random polypeptide of 100 amino acids, the configurational entropy, S_{cr}, may be calculated using eq. 8-2c and eq. 8-7 as follows:

$$S_{cr} = k \ln \Omega_{cr}$$

$$\text{since } \Omega_{cr} = \frac{N!}{n_1!n_2!...n_{20}!} = \frac{100!}{5!\,5!....5!} = \frac{100!}{(5!)^{20}}$$

$$= 1.28 \times 10^{115} \qquad (8\text{-}8)$$

The calculation of equation 8-8 assumes that an equal number of each type of amino acid, namely 5, are contained in the polypeptide. Since k, or Boltzmann's constant, equals 1.38×10^{-16} erg/deg, and ln $(1.28 \times 10^{115}) = 265$,

$$S_{cr} = 1.38 \times 10^{-16} \times 265 = 3.66 \times 10^{-14} \text{ erg/deg-polypeptide}$$

If only *one* specific sequence of amino acids could give the proper function, then the configurational entropy for the protein or specified, aperiodic polypeptide would be given by

$$\begin{aligned} S_{cm} &= k \ln \Omega_{cm} \\ &= k \ln 1 \\ &= 0 \end{aligned} \qquad (8\text{-}9)$$

Determining ΔS_c in Going from a Random Polymer to an Informed Polymer

The change in configurational entropy, ΔS_c, as one goes from a random polypeptide of 100 amino acids with an equal number of each amino acid type to a polypeptide with a specific message or sequence is:

$$\Delta S_c = S_{cm} - S_{cr}$$

$$= 0 - 3.66 \times 10^{-14} \text{ erg/deg-polypeptide}$$

$$= -3.66 \times 10^{-14} \text{ erg/deg-polypeptide} \qquad (8\text{-}10)$$

The configurational entropy work ($-T\Delta S_c$) at ambient temperatures is given by

$$-T\Delta S_c = -(298K) \times (-3.66 \times 10^{-14}) \text{ erg/deg-polypeptide}$$

$$= 1.1 \times 10^{-11} \text{ erg/polypeptide}$$

$$= 1.1 \times 10^{-11} \text{ erg/polypeptide} \times \frac{6.023 \times 10^{23} \text{ molecules/mole}}{10{,}000 \text{ gms/mole}} \times \frac{1 \text{ cal}}{4.184 \times 10^{7} \text{ ergs}}$$

$$= 15.8 \text{ cal/gm} \qquad (8\text{-}11)$$

where the protein mass of 10,000 amu was estimated by assuming an average amino acid weight of 100 amu after the removal of the water molecule. Determination of the configurational entropy work for a protein containing 300 amino acids equally divided among the twenty types gives a similar result of 16.8 cal/gm.

In like manner the configurational entropy work for a DNA molecule such as for *E. coli* bacterium may be calculated assuming 4×10^6 nucleotides in the chain with 1×10^6 each of the four distinctive nucleotides, each distinguished by the type of base attached, and each nucleotide assumed to have an average mass of 339 amu. At 298K:

$$-T\Delta S_c = -T(S_{cm} - S_{cr})$$

$$= T(S_{cr} - S_{cm})$$

$$= kT \ln (\Omega_{cr} - \ln \Omega_{cm})$$

$$= kT \ln \left[\frac{(4 \times 10^6)!}{(10^6)!\,(10^6)!\,(10^6)!\,(10^6)!} \right] - kT \ln 1$$

$$= 2.26 \times 10^{-7} \text{ erg/polynucleotide}$$

$$= 2.39 \text{ cal/gm} \qquad (8\text{-}12)$$

It is interesting to note that, while the work to code the DNA molecule with 4 million nucleotides is much greater than the work required to code a protein of 100 amino acids (2.26×10^{-7} erg/DNA vs. 1.10×10^{-11} erg/protein), the work per gram to code such molecules is cules is actually less in DNA. There are two reasons for this perhaps unexpected result: first, the nucleotide is more massive than the amino acid (339 amu vs. 100 amu); and second, the alphabet is more limited, with only four useful nucleotide "letters" as compared to twenty useful amino acid letters. Nevertheless, it is the total work that is important, which means that synthesizing DNA is much more difficult than synthesizing protein.

It should be emphasized that these estimates of the magnitude of the configurational entropy work required are conservatively small. As a practical matter, our calculations have ignored the configurational entropy work involved in the selection of monomers. Thus, we have assumed that only the proper subset of 20 biologically significant amino acids was available in a prebiotic oceanic soup to form a biofunctional protein. The same is true of DNA. We have assumed that in the soup only the proper subset of 4 nucleotides was present and that these nucleotides do not interact with amino acids or other soup ingredients. As we discussed in Chapter 4, many varieties of amino acids and nucleotides would have been present in a real ocean—varieties which have been ignored in our calculations of configurational entropy work. In addition, the soup would have contained many other kinds of molecules which could have reacted with amino acids and nucleotides. The problem of using only the appropriate optical isomer has also been ignored. A random chemical soup would have contained a 50-50 mixture of D- and L-amino acids, from which a true protein could incorporate only the L-enantiomer. Similarly, DNA uses exclusively the optically active sugar D-deoxyribose. Finally, we have ignored the problem of forming unnatural links, assuming for the calculations that only α-links occurred between amino acids in making polypeptides, and that only correct linking at the 3′,5′-position of sugar occurred in forming polynucleotides. A quantification of these problems of specificity has recently been made by Yockey.[21]

The dual problem of *selecting* the proper composition of matter and then *coding* or rearranging it into the proper sequence is analogous to writing a story using letters drawn from a pot containing many duplicates of each of the 22 Hebrew consonants and 24 Greek and 26 English letters all mixed together. To write in English the message

HOW DID I GET HERE?

we must first draw from the pot 2 Hs, 2 Is, 3 Es, 2 Ds, and one each of the letters W, O, G, T, and R. Drawing or selecting this specific set of letters would be a most unlikely event itself. The work of selecting just these 14 letters would certainly be far greater than arranging them in the correct sequence. Our calculations only considered the easier step of coding while ignoring the greater problem of selecting the correct set of letters to be coded. We thereby greatly underestimate the actual configurational entropy work to be done.

In Chapter 6 we developed a scale showing degrees of investigator interference in prebiotic simulation experiments. In discussing this scale it was noted that very often in reported experiments the experimenter has actually played a crucial but *illegitimate* role in the success of the experiment. It becomes clear at this point that one illegitimate role of the investigator is that of providing a portion of the configurational entropy work, i.e., the "selecting" work portion of the total $-T\,\Delta S_c$ work.

It is sometimes argued that the type of amino acid that is present in a protein is critical only at certain positions—active sites—along the chain, but not at every position. If this is so, it means the same message (i.e., function) can be produced with more than one sequence of amino acids.

This would reduce the coding work by making the number of permissible arrangements Ω_{cm} in eqs. 8-9 and 8-10 for S_{cm} greater than 1. The effect of overlooking this in our calculations, however, would be negligible compared to the effect of overlooking the "selecting" work and only considering the "coding" work, as previously discussed. So we are led to the conclusion that our estimate for ΔS_c is very conservatively low.

Calculating the Total Work: Polymerization of Biomacromolecules

It is now possible to estimate the total work required to combine biomonomers into the appropriate polymers essential to living systems. This calculation using eq. 8-5 might be thought of as occurring in two steps. First, amino acids polymerize into a polypeptide, with the chemical and thermal entropy work being accomplished ($\Delta H - T\,\Delta S_{th}$). Next, the random polymer is rearranged into a specific sequence which constitutes doing configurational entropy work

($-T\Delta S_c$). For example, the total work as expressed by the change in Gibbs free energy to make a specified sequence is

$$\Delta G = \Delta H - T\Delta S_{th} - T\Delta S_c \qquad (8\text{-}13)$$

where $\Delta H - T\Delta S_{th}$ may be assumed to be 300 kcal/mole to form a random polypeptide of 101 amino acids (100 links). The work to code this random polypeptide into a useful sequence so that it may function as a protein involves the additional component of $-T\Delta S_c$ "coding" work, which has been estimated previously to be 15.9 cal/gm, or approximately 159 kcal/mole for our protein of 100 links with an estimated mass of 10,000 amu per mole. Thus, the total work (neglecting the "sorting and selecting" work) is approximately

$$\Delta G = (300 + 159) \text{ kcal/mole} = 459 \text{ kcal/mole} \qquad (8\text{-}14)$$

with the coding work representing 159/459 or 35% of the total work.

In a similar way, the polymerization of 4×10^6 nucleotides into a random polynucleotide would require approximately 27×10^6 kcal/mole. The coding of this random polynucleotide into the specified, aperiodic sequence of a DNA molecule would require an additional 3.2×10^6 kcal/mole of work. Thus, the fraction of the total work that is required to code the polymerized DNA is seen to be 8.5%, again neglecting the "sorting and selecting" work.

The Impossibility of Protein Formation under Equilibrium Conditions

It was noted in Chapter 7 that because macromolecule formation (such as amino acids polymerizing to form protein) goes uphill energetically, work must be done on the system via energy flow through the system. We can readily see the difficulty in getting polymerization reactions to occur under equilibrium conditions, i.e., in the absence of such an energy flow.

Under equilibrium conditions the concentration of protein one would obtain from a solution of 1 M concentration in each amino acid is given by:

$$K = \frac{[\text{protein}] \times [H_2O]}{[\text{glycine}]\,[\text{alanine}]\ldots} \qquad (8\text{-}15)$$

where K is the equilibrium constant and is calculated by

$$K = \exp\left(\frac{-\Delta G}{RT}\right) \tag{8-16}$$

An equivalent form is

$$\Delta G = -RT \ln K \tag{8-17}$$

We noted earlier that $\Delta G = 459$ kcal/mole for our protein of 101 amino acids. The gas constant R = 1.9872 cal/deg-mole and T is assumed to be 298K. Substituting these values into eqs. 8-15 and 8-16 gives

$$\text{protein concentration} = 10^{-338}\text{M} \tag{8-18}$$

This trivial yield emphasizes the futility of protein formation under equilibrium conditions. In the next chapter we will consider various theoretical models attempting to show how energy flow through the system can be useful in doing the work quantified in this chapter for the polymerization of DNA and protein. Finally, we will examine experimental efforts to accomplish biomacromolecule synthesis.

References

1. Peter M. Molton, 1978. *J. Brit. Interplanet. Soc.* **31**, 147.
2. G. Nicolis and I. Prigogine, 1977. *Self Organization in Nonequilibrium Systems*. New York: John Wiley, p. 25.
3. I. Prigogine, G. Nicolis, and A. Babloyantz, 1972. *Physics Today*, p. 23.
4. L.E. Orgel, 1973. *The Origins of Life*. New York: John Wiley, p. 189ff; M. Polanyi, 1968. *Science* **160**, 1308; Hubert P. Yockey, 1977. *J. Theoret. Biol.* **67**, 377; Jeffrey Wickens, 1978. *J. Theoret. Biol.* **72**, 191.
5. Yockey, *J. Theoret. Biol.*, p. 383.
6. Orgel, *The Origins of Life*, p. 189.
7. Yockey, *J. Theoret. Biol.*, p. 579.
8. Wickens, *J. Theoret. Biol.*, p. 191.
9. Orgel, *The Origins of Life*, p. 190.
10. L. Brillouin, 1951. *J. Appl. Phys.* **22**, 334; 1951. *J. Appl. Phys.* **22**, 338; 1950. *Amer. Sci.* **38**, 594; 1949. *Amer. Sci.* **37**, 554.

11. E. Schrodinger, 1945. *What is Life?* London: Cambridge University Press, and New York: Macmillan.
12. W. Ehrenberg, 1967. *Sci. Amer.* **217**, 103; Myron Tribus and Edward C. McIrvine, 1971. *Sci. Amer.* **225**, 197.
13. Brillouin, *J. Appl. Phys.* **22**, 335.
14. John O. Hutchens, 1976. *Handbook of Biochemistry and Molecular Biology*, 3rd ed., Physical and Chemical Data, Gerald D. Fasman. Cleveland: CRC Press.
15. H. Morowitz, 1968. *Energy Flow in Biology*. New York: Academic Press, p. 79.
16. H. Borsook and H.M. Huffman, 1944. *Chemistry of Amino Acids and Proteins*, ed. C.L.A. Schmidt. Springfield, Mass.: Charles C. Thomas Co., p. 822.
17. Wickens, *J. Theoret. Biol.*, p. 191.
18. Morowitz, *Energy Flow in Biology*, p. 79.
19. D.W. Armstrong, F. Nome, J.H. Fendler, and J. Nagyvary, 1977. *J. Mol. Evol.* **9**, 213.
20. Brillouin, *J. Appl. Phys.* **22**, 338.
21. H.P. Yockey, 1981. *J. Theoret. Biol.* **91**, 13.

CHAPTER 9

Specifying How Work Is To Be Done

In Chapter 7 we saw that the work necessary to polymerize DNA and protein molecules from simple biomonomers could *potentially* be accomplished by energy flow through the system. Still, we know that such energy flow is a necessary but not sufficient condition for polymerization of the macromolecules of life. Arranging a pile of bricks into the configuration of a house requires work. One would hardly expect to accomplish this work with dynamite, however. Not only must energy flow through the system, it must be coupled in some specific way to the work to be done. This being so, we devoted Chapter 8 to identifying various components of work in typical polymerization reactions. In reviewing those individual work components, one thing became clear. The coupling of energy flow to the specific work requirements in the formation of DNA and protein is particularly important since the required configurational entropy work of coding is substantial.

Theoretical Models for the Origin of DNA and Protein

A mere appeal to open system thermodynamics does little good. What must be done is to advance a workable theoretical model of *how* the available energy can be coupled to do the required work. In this chapter various theoretical models for the origin of DNA and protein will be evaluated. Specifically, we will discuss how each model proposes to couple the available energy to the required work, particularly the configurational entropy work of coding.

Chance

Before the specified complexity of living systems began to be appreciated, it was thought that, given enough time, "chance" would explain the origin of living systems. In fact, most textbooks state that chance is the basic explanation for the origin of life. For example, Lehninger in his classic textbook *Biochemistry* states,

> We now come to the critical moment in evolution in which the first semblance of "life" appeared, through the chance association of a number of abiotically formed macromolecular components, to yield a unique system of greatly enhanced survival value.[1]

More recently the viability of "chance" as a mechanism for the origin of life has been severely challenged.[2]

We are now ready to analyze the "chance" origin of life using the approach developed in the last chapter. This view usually assumes that energy flow through the system is capable of doing the chemical and the thermal entropy work, while the configurational entropy work of both selecting and coding is the fortuitous product of chance.

To illustrate, assume that we are trying to synthesize a protein containing 101 amino acids. In eq. 8-14 we estimated that the total free energy increase (ΔG) or work required to make a random polypeptide from previously selected amino acids was 300 kcal/mole. An additional 159 kcal/mole is needed to code the polypeptide into a protein. Since the "chance" model assumes no coupling between energy flow and sequencing, the fraction of the polypeptide that has the correct sequence may be calculated (eq. 8-16) using equilibrium thermodynamics; i.e..

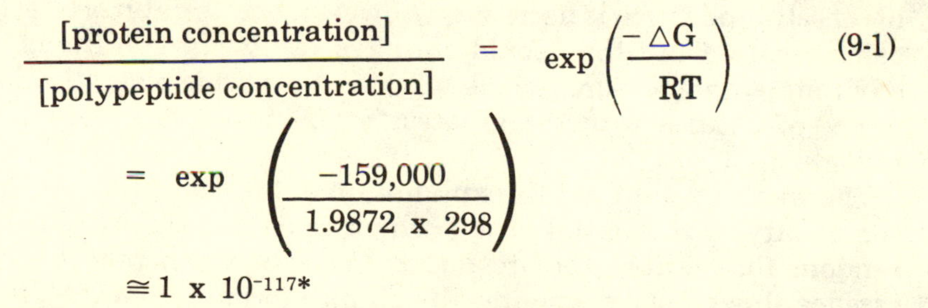

$$\frac{[\text{protein concentration}]}{[\text{polypeptide concentration}]} = \exp\left(\frac{-\Delta G}{RT}\right) \qquad (9\text{-}1)$$

$$= \exp\left(\frac{-159{,}000}{1.9872 \times 298}\right)$$

$$\cong 1 \times 10^{-117}\text{*}$$

*This is essentially the inverse of the estimate for the number of ways one can arrange 101 amino acids in a sequence (i.e., I/Ω_c in eq. 8-7).

This ratio gives the fraction of polypeptides that have the right sequence to be a protein.

Eigen[3] has estimated the number of polypeptides of molecular weight 10^4 (the same weight used in our earlier calculations) that would be found in a layer 1 meter thick covering the surface of the entire earth. He found it to be 10^{41}. If these polypeptides reformed with new sequences at the maximum rate at which chemical reactions may occur, namely 10^{14}/s, for 5×10^9 years (1.6×10^{17}s), the total number of polypeptides that would be formed during the assumed history of the earth would be

$$10^{41} \times 10^{14}/\text{s} \times 1.6 \times 10^{17}\text{s} = 10^{72} \qquad (9\text{-}2)$$

Combining the results of eq. 9-1 and 9-2, we find the probability of producing one protein of 101 amino acids in five billion years is only $1/10^{45}$. Using somewhat different illustrations, Steinman[4] and Cairns-Smith[5] also come to the conclusion that chance is insufficient.

It is apparent that "chance" should be abandoned as an acceptable model for coding of the macromolecules essential in living systems. In fact, it has been, except in introductory texts and popularizations.

Neo-Darwinian Natural Selection

The widespread recognition of the severe improbability that self-replicating organisms could have formed from purely random interactions has led to a great deal of speculation—speculation that some organizing principle must have been involved. In the company of many others, Crick[6] has considered that the neo-Darwinian mechanism of natural selection might provide the answer. An entity capable of self-replication is necessary, however, before natural selection can operate. Only then could changes result via mutations and environmental pressures which might in turn bring about the dominance of entities with the greatest probabilities of survival and reproduction.

The weakest point in this explanation of life's origin is the great complexity of the initial entity which must form, apparently by random fluctuations, before natural selection can take over. In essence this theory postulates the chance formation of the "metabolic motor" which will subsequently be capable of channeling energy flow through the system. Thus harnessed by coupling through the metabolic motor, the energy flow is imagined to supply not only chemical and thermal entropy work, but also the configurational entropy work of selecting the appropriate chemicals and then coding the resultant polymer into an aperiodic, specified, biofunctioning polymer. As a minimum, this system must carry in its structure the information for its own synthesis, and control the machinery which will fabricate any desired copy. It is widely agreed that such a system requires both protein and nucleic acid.[7] This view is not unanimous, however. A few have suggested that a short peptide would be sufficient.[8]

One way out of the problem would be to extend the concept of natural selection to the pre-living world of molecules. A number of authors have entertained this possibility, although no reasonable explanation has made the suggestion plausible. Natural selection is a recognized principle of differential reproduction which presupposes the existence of at least two distinct types of self-replicating molecules. Dobzhansky appealed to those doing origin-of-life research not to tamper with the definition of natural selection when he said:

> I would like to plead with you, simply, please realize you cannot use the words "natural selection" loosely. Prebiological natural selection is a contradiction in terms.[9]

Bertalanffy made the point even more cogently:

> Selection, i.e., favored survival of "better" precursors of life, already presupposes self-maintaining, complex, open systems which may compete; therefore selection cannot account for the origin of such systems.[10]

Inherent Self-Ordering Tendencies in Matter

How could energy flow through the system be sufficiently coupled to do the chemical and thermal entropy work to form a nontrivial yield of polypeptides (as previously assumed in the "chance" model)? One answer has been the suggestion that configurational entropy work, especially the coding work, could occur as a consequence of the self-ordering tendencies in matter. The experimental work of Steinman and Cole[11] in the late Sixties is still widely cited in support of this model.[12] The polymerization of protein is hypothesized to be a *nonrandom* process, the coding of the protein resulting from differences in the chemical bonding forces. For example, if amino acids A and B react chemically with one another more readily than with amino acids C, D, and E, we should expect to see a greater frequency of AB peptide bonds in protein than AC, AD, AE, or BC, BD, BE bonds.

Together with our colleague Randall Kok, we have recently analyzed the ten proteins originally analyzed by Steinman and Cole,[13] as well as fifteen additional proteins whose structures (except for hemoglobin) have been determined since their work was first published in 1967. Our expectation in this study was that one would only get agreement between the dipeptide bond frequencies from Steinman and Cole's work and those observed in actual proteins if one considered a large number of proteins averaged together. The distinctive structures of individual proteins would cause them to vary greatly from Steinman and Cole's data, so only when these distinctives are averaged out could one expect to approach Steinman and Cole's dipeptide bond frequency results. The reduced data presented in table 9-1 shows that Steinman and Cole's dipeptide bond frequencies do not correlate well with the observed peptide bond frequencies for one, ten, or twenty-five proteins. It is a simple matter to make such calculations on an electronic digital computer. We surmise that additional assumptions not stated in their paper were used to achieve the better agreements.

Furthermore, the peptide bond frequencies for the twenty-five proteins approach a distribution predicted by random statistics rather than the dipeptide bond frequency measured by Steinman and Cole. This observation means that bonding preferences between various amino acids play no significant role in coding protein. Finally, if chemical bonding forces were influential in amino acid sequencing, one would expect to get a single sequence (as in ice crystals) or no more than a few sequences, instead of the large variety we observe in living systems. Yockey, with a different analysis, comes to essentially the same conclusion.[14]

A similar conclusion may be drawn for DNA synthesis. No one to date has published data indicating that bonding preferences could have had any role in coding the DNA molecules. Chemical bonding forces apparently have minimal effect on the sequence of nucleotides in a polynucleotide.

Table 9-1.
Comparison of Steinman and Cole's experimentally determined dipeptide bond frequencies, and frequencies calculated by Steinman and Cole, and by Kok and Bradley from known protein sequences.

Dipeptide*	Values (relative to Gly-Gly)			
	S/C+		K/B#	
	exp‡	cal	cal-wa	cal-woa
Gly-Gly	1.0	1.0	1.0 (1.0) [1.0]	1.0 (1.0) [1.0]
Gly-Ala	0.8	0.7	1.1 (1.1) [2.0]	2.0 (1.2) [1.0]
Ala-Gly	0.8	0.6	1.0 (1.1) [2.2]	1.5 (1.2) [0.0]
Ala-Ala	0.7	0.6	1.3 (1.5) [4.4]	2.8 (1.5) [0.0]
Gly-Val	0.5	0.2	0.2 (0.3) [0.4]	1.5 (1.2) [1.0]
Val-Gly	0.5	0.3	0.3 (0.3) [0.6]	0.8 (0.6) [0.0]
Gly-Leu	0.5	0.3	0.3 (0.3) [0.2]	1.3 (0.7) [1.0]
Leu-Gly	0.5	0.2	0.3 (0.3) [0.8]	1.3 (1.0) [1.0]
Gly-Ile	0.3	0.1	0.1 (0.2) [0.6]	1.0 (0.8) [0.0]
Ile-Gly	0.3	0.1	0.1 (0.2) [0.2]	0.0 (0.4) [0.0]
Gly-Phe	0.1	0.1	0.1 (0.2) [0.4]	0.5 (0.5) [0.0]
Phe-Gly	0.1	0.1	0.1 (0.1) [0.6]	1.0 (0.5) [1.0]

(Adapted after G. Steinman and M.V. Cole, 1967. *Proc. Nat. Acad. Sci. U.S.* **58**, 735).

*The dipeptides are listed in terms of increasing volume of the side chains of the constituent residues. Gly = glycine, Ala = alanine, Val = valine, Leu = leucine, Ile = isoleucine and Phe = phenylalanine. Example: Gly-Ala = glycylalanine.

+Steinman and Cole's (S/C) experimentally determined dipeptide bond frequencies were normalized and compared to the calculated frequencies obtained by counting actual peptide bond frequencies in ten proteins, assuming all seryl and threonyl residues are counted as glycine and all aspartyl and glutamyl residues are counted as alanine. The ten proteins used were: egg lysozyme, ribonuclease, sheep insulin, whale myoglobin, yeast cytochrome c, tobacco mosaic virus, β-corticotropin, glucagon, melanocyte-stimulating hormone, and chymotrypsinogen. Because of ambiguity regarding sequences used by S/C, all sequences are those shown in *Atlas of Protein Sequence and Structure*, 1972. Vol. V (ed. by M.O. Dayhoff). National Biomedical Research Foundation, Georgetown University Medical Center, Washington, D.C.

‡The experimentally determined dipeptide frequencies were obtained with aqueous solutions containing 0.01 M each amino acid, 0.125 N HCl, 0.1 M sodium dicyanamide.

#Kok and Bradley's (K/B) calculated dipeptide frequencies were obtained by counting actual peptide bond frequencies for the same ten proteins with (wa) and without (woa) S/C assumptions. The numbers in brackets are for one protein, enterotoxin B, with (wa) and without (woa) S/C assumptions. The numbers in parentheses are for twenty-five proteins with (wa) and without (woa) S/C assumptions. The twenty-five proteins are the ten used by S/C and alpha S1 Casein (bovine); azurin (bordetella bronchisep-

tica); carboxypeptidase A (bovine); cytochrome b5 (bovine); enterotoxin B; elastase (pig); glyceraldehyde 3-phosphate dehydrogenase (lobster); human growth hormone; human hemoglobin beta chain; histone IIB2 (bovine); immunoglobulin gamma-chain 1, V-I (human EU); penicillinase (bacillus licheniformis 749/c); sheep prolactin; subtilisin (bacillus amyloliquefaciens); and tryptophan synthetase alpha chain (*E-coli* K-12). Sequences are those shown in *Atlas of Protein Sequence and Structure*, 1972. Vol. V (ed. by M.O. Dayhoff). Note disagreement between S/C and K/B calculated results. Also K/B calculated results are at variance with S/C experimental values for one, ten or twenty-five proteins, with (wa) or without (woa) S/C assumptions.

Mineral Catalysis

Mineral catalysis is often suggested as being significant in prebiotic evolution. In the experimental investigations reported in the early 1970s,[15] mineral catalysis in polymerization reactions was found to operate by adsorption of biomonomers on the surface or between layers of clay. Monomers were effectively concentrated and protected from rehydration so that condensation polymerization could occur. There does not appear to be any additional effect. In considering this catalytic effect of clay, Hulett has advised, "It must be remembered that the surface cannot change the free energy relationships between reactants and products, but only the speed with which equilibrium is reached."[16]

Is mineral catalysis capable of doing the chemical work and/or thermal entropy work? The answer is a qualified no. While it should assist in doing the thermal entropy work, it is incapable of doing the chemical work since clays do not supply energy. This is why successful mineral catalysis experiments invariably use energy-rich precursors such as aminoacyl adenylates rather than amino acids.[17]

Is there a real prospect that mineral catalysis may somehow accomplish the configurational entropy work, particularly the coding of polypeptides or polynucleotides? Here the answer is clearly no. In all experimental work to date, only random polymers have been condensed from solutions of selected ingredients. Furthermore, there is no theoretical basis for the notion that mineral catalysis could impart any significant degree of information content to polypeptides or polynucleotides. As has been noted by Wilder-Smith,[18] there is really no reason to expect the low-grade order resident on minerals to impart any high degree of coding to polymers that condense while adsorbed on the mineral's surface. To put it another way, one cannot get a complex, aperiodic-sequenced polymer using a very periodic (or crystalline) template.

In summary, mineral catalysis must be rejected as a mechanism for doing either the chemical or configurational entropy work required to polymerize the macromolecules of life. It can only assist in polymerizing short, random chains of polymers from selected high-energy biomonomers by assisting in doing the thermal entropy work.

Nonlinear, Nonequilibrium Processes

1. Ilya Prigogine

Prigogine has developed a more general formulation of the laws of thermodynamics which includes nonlinear, irreversible processes such as autocatalytic activity. In his book *Self Organization in Nonequilibrium Systems* (1977)[19] co-authored with Nicolis, he summarized this work and its application to the organization and maintenance of highly complex structures in living things. The basic thesis in the book is that there are some systems which obey nonlinear laws—laws that produce two distinct kinds of behavior. In the neighborhood of thermodynamic equilibrium, destruction of order prevails (entropy achieves a maximum value consistent with the system constraints). If these same systems are driven sufficiently far from equilibrium, however, ordering may appear spontaneously.

Heat flow by convection is an example of this type of behavior. Heat conduction in gases normally occurs by the random collision of gas molecules. Under certain conditions, however, heat conduction may occur by a heat-convection current—the coordinated movement of many gas molecules. In a similar way, water flow out of a bathtub may occur by random movement of the water molecules under the influence of gravity. Under certain conditions, however, this random movement of water down the drain is replaced by the familiar soapy swirl—the highly coordinated flow of the vortex. In each case random movements of molecules in a fluid are spontaneously replaced by a highly ordered behavior. Prigogine et al.,[20] Eigen,[21] and others have suggested that a similar sort of self-organization may be intrinsic in organic chemistry and can potentially account for the highly complex macromolecules essential for living systems.

But such analogies have scant relevance to the origin-of-life question. A major reason is that they fail to distinguish between order and complexity. The highly ordered movement of energy through a system as in convection or vortices suffers from the same shortcoming as the analogies to the static, periodic order of crystals. Regularity or order cannot serve to store the large amount of information required by living systems. A highly irregular, but specified, structure is required rather than an ordered structure. This is a serious flaw in the analogy offered. There is no apparent connection between the kind of spontaneous ordering that occurs from energy flow through such systems and the work required to build aperiodic information-intensive macromolecules like DNA and protein. Prigogine et al.[22] suggest that the energy flow through the system decreases the system entropy, leading potentially to the highly organized structure of DNA and protein. Yet they offer no suggestion as to how the decrease in thermal entropy from energy flow through the system could be coupled to do the configurational entropy work required.

A second reason for skepticism about the relevance of the models developed by Prigogine et al.,[23] and others is that ordering produced within the system arises through constraints imposed in an implicit way at the system boundary. Thus, the system order, and more importantly the system complexity, cannot exceed that of the environment.

Walton[24] illustrates this concept in the following way. A container of gas placed in contact with a heat source on one side and a heat sink on the opposite side is an open system. The flow of energy through the system from the heat source to the heat sink forms a concentration relative to the gas in the cooler region. The order in this system is established by the structure: source-intermediate systems-sink. If this structure is removed, allowing the heat source to come into contact with the heat sink, the system decays back to equilibrium. We should note that the information induced in an open system doesn't exceed the amount of information built into the structural environment, which is its source.

Condensation of nucleotides to give polynucleotides or nucleic acids can be brought about with the appropriate apparatus (i.e., structure) and supplies of energy and matter. Just as in Walton's illustration, however, Mora[25] has shown that the amount of order (not to mention specified complexity) in the final product is no greater than the amount of information introduced in the physical structure of the experiment or chemical structure of the reactants. Non-equilibrium thermodynamics does not account for this structure, but assumes it and then shows the kind of organization which it produces. The origin and maintenance of the structure are not explained, and as Harrison[26] correctly notes, this question leads back to the origin of structure in the universe. Science offers us no satisfactory answer to this problem at present.

Nicolis and Prigogine[27] offer their trimolecular model as an example of a chemical system with the required nonlinearity to produce self ordering. They are able to demonstrate mathematically that within a system that was initially homogeneous, one may subsequently have a periodic, spatial variation of concentration. To achieve this low degree of ordering, however, they must require boundary conditions that could only be met at cell walls (i.e., at membranes), relative reaction rates that are atypical of those observed in condensation reactions, a rapid removal of reaction products, and a trimolecular reaction (the highly unlikely simultaneous collision of three atoms). Furthermore the trimolecular model requires chemical reactions that are essentially irreversible. But condensation reactions for polypeptides or polynucleotides are highly reversible unless all water is removed from the system.

They speculate that the low degree of spatial ordering achieved in the simple trimolecular model could potentially be orders of magnitude greater for the more complex reactions one might observe leading up to a fully replicating cell. The list of boundary constraints, relative reaction rates, etc. would, however, also be orders of magnitude larger. As a matter of fact, one is left with so constraining the system at the boundaries that ordering is inevitable from the structuring of the environment by the chemist. The fortuitous satisfaction of all of these boundary constraints simultaneously would be a miracle in its own right.

It is possible at present to synthesize a few proteins such as insulin in the laborabory. The chemist supplies not only energy to do the chemical and thermal entropy work, however, but also the necessary chemical manipulations to accomplish the configurational entropy work. Without this, the selection of the proper composition and the coding for the right sequence of amino acids would not occur. The success of the experiment is fundamentally dependent on the chemist.

Finally, Nicolis and Prigogine have postulated that a system of chemical reactions which explicitly shows autocatalytic activity may ultimately be able to circumvent the problems now associated with synthesis of prebiotic DNA and protein. It remains to be demonstrated *experimentally*, however, that these models have any real correspondence to prebiotic condensation reactions. At best, these models predict higher yields without any mechanism to control sequencing. Accordingly, no experimental evidence has been reported to show how such models could have produced any significant degree of coding. No, the models of Prigogine et al., based on nonequilibrium thermodynamics, do not at present offer an explanation as to how the configurational entropy work is accomplished under prebiotic conditions. The problem of how to couple energy flow through the system to do the required configurational entropy work remains.

2. Manfred Eigen

In his comprehensive application of nonequilibrium thermodynamics to the evolution of biological systems, Eigen[28] has shown that selection could produce no evolutionary development in an open system unless the system were maintained far from equilibrium. The reaction must be autocatalytic but capable of self-replication. He develops an argument to show that in order to produce a truly self-replicating system the complementary base-pairing instruction potential of nucleic acids must be combined with the catalytic coupling function of proteins. Kaplan[29] has suggested a minimum of 20-40 functional proteins of 70-100 amino acids each, and a similar number of nucleic acids would be required by such a system. Yet as has previously been noted, the chance origin of even one protein of 100 amino acids is essentially zero.

The shortcoming of this model is the same as for those previously discussed; namely, no way is presented to couple the energy flow through the system to achieve the configurational entropy work required to create a system capable of replicating itself.

Periodically we see reversions (perhaps inadvertent ones) to chance in the theoretical models advanced to solve the problem. Eigen's model illustrates this well. The model he sets forth must necessarily arise from chance events and is nearly as incredible as the chance origin of life itself. The fact that generally chance has to be invoked many times in the abiotic sequence has been called by Brooks and Shaw "a major weakness in the whole chemical evolutionary theory."[30]

Experimental Results in Synthesis of Protein and DNA

Thus far we have reviewed the various theoretical models proposed to explain how energy flow through a system might accomplish the work of synthesizing protein and DNA macromolecules, but found them wanting. Nevertheless, it is conceivable that experimental support for a spontaneous origin of life can be found in advance of the theoretical explanation for how this occurs. What then can be said of the experimental efforts to synthesize protein and DNA macromolecules? Experimental efforts to this end have been enthusiastically pursued for the past thirty years. In this section, we will review efforts toward the prebiotic synthesis of both protein and DNA, considering the three forms of energy flow most commonly thought to have been available on the early earth. These are thermal energy (volcanoes), radiant energy (sun), and chemical energy in the form of either condensing agents or energy-rich precursors. (Electrical energy is excluded at this stage of evolution as being too "violent," destroying rather than joining the biomonomers.)

Thermal Synthesis

Sidney Fox[31] has pioneered the thermal synthesis of polypeptides, naming the products of his synthesis *proteinoids*. Beginning with either an aqueous solution of amino acids or dry ones, he heats his material at 200° C* for 6-7 hours. All initial solvent water, plus water produced during polymerization, is effectively eliminated through vaporization. This elimination of the water makes possible a small but significant yield of polypeptides, some with as many as 200 amino acid units. Heat is introduced into the system by conduction and convection and leaves in the form of steam. The reason for the success of the polypeptide formation is readily seen by examining again equations 8-15 and 8-16. Note that increasing the temperature would increase the product yield through increasing the value of exp $(-\Delta G/RT)$. But more importantly, eliminating the water makes the reaction irreversible, giving an enormous increase in yield over that observed under equilibrium conditions by the application of the law of mass action.

Thermal syntheses of polypeptides fail, however, for at least four reasons. First, studies using nuclear magnetic resonance (NMR) have shown that thermal proteinoids "have scarce resemblance to natural peptidic material because β, γ, and ϵ peptide bonds largely

*Fox has modified this picture in recent years by developing "low temperature" syntheses, i.e., 90°-120° C. See S. Fox, 1976. *J. Mol. Evol.* **8**, 301; and D. Rohlfing, 1976. *Science* **193**, 68.

predominate over α-peptide bonds."*[32] Second, thermal proteinoids are composed of approximately equal numbers of L- and D-amino acids in contrast to viable proteins with all L-amino acids. Third, there is no evidence that proteinoids differ significantly from a random sequence† of amino acids, with little or no catalytic activity. Miller and Orgel have made the following observation with regard to Fox's claim that proteinoids resemble proteins:

> The degree of nonrandomness in thermal polypeptides so far demonstrated is minute compared to nonrandomness of proteins. It is deceptive, then, to suggest that thermal polypeptides are similar to proteins in their nonrandomness.[33]

Fourth, the geological conditions indicated are too unreasonable to be taken seriously. As Folsome has commented, "The central question [concerning Fox's proteinoids] is where did all those pure, dry, concentrated, and optically active amino acids come from in the real, abiological world?"[34]

There is no question that thermal energy flow through the system including the removal of water is accomplishing the thermal entropy and chemical work required to form a polypeptide (300 kcal/mole in our earlier example). The fact that polypeptides are formed is evidence of the work done. It is equally clear that the additional configurational entropy work required to convert an aperiodic unspecified polypeptide into a specified, aperiodic polypeptide which is a functional protein has not been done (159 kcal/mole in our earlier example).

It should be remembered that this 159 kcal/mole of configurational entropy work was calculated assuming the sequencing of the amino acids was the only additional work to be done. Yet the experimental results of Temussi et al.,[35] indicate that obtaining all L-amino acids from a racemic mixture and getting α-linking between the amino acids are quite difficult. This requirement further increases the configurational entropy work needed over that estimated to do the coding work (159 kcal/mole). We may estimate the magnitude of this increase in the configurational entropy work term by returning to our original calculations (eq. 8-7 and 8-8).

In our original calculation for a hypothetical protein of 100 amino acid units, we assumed the amino acids were equally divided among

---

*This quotation refers to peptide links involving the β-carboxyl group of aspartic acid, the γ-carboxyl group of glutamic acid, and the ϵ-amino group of lysine which are never found in natural proteins. Natural proteins use α-peptide bonds exclusively.

†It is noted, however, that Fox has long disputed this.

the twenty types. We calculated the number of possible amino acid sequences as follows:

$$\Omega_{cr} = \frac{100!}{5!5!5!...5!} = \frac{100!}{(5!)^{20}} = 1.28 \times 10^{115} \quad (9\text{-}3)$$

If we note that at each site the probability of having an L-amino acid is 50%, and make the generous assumption* that there is a 50% probability that a given link will be of the α-type observed in true proteins, then the number of ways the system can be arranged in a random chemical reaction is given by

$$\Omega_{cr} = 1.28 \times 10^{115} \times 2^{100} \times 2^{99} = 10^{175} \quad (9\text{-}4)$$

where 2^{100} refers to the number of additional arrangements possible, given that each site could contain an L- or D-amino acid, and 2^{99} assumes the 99 links between the 100 amino acids in general are equally divided between the natural α-links and the unnatural β-, γ-, or ϵ-links.

The requirements for a biologically functional protein molecule are: (1) all L-amino acids, (2) all α-links, and (3) a specified sequence. This being so, the calculation of the configurational entropy of the protein molecule using equation 8-8 is unchanged except that the number of ways the system can be arranged, Ω_{cr}, is increased from 1.28×10^{115} to 1.0×10^{175} as shown in equations 9-3 and 9-4. We may use the relationships of equations 8-7 and 8-8 but with the number of permutations modified as shown here to find a total configurational entropy work. When we do, we get a total configurational entropy work of 195 kcal/mole, of which 159 kcal/mole is for sequencing and 36 kcal/mole to attain all L-amino acids and all α-links. Finally, it should be recognized that Fox and others who use his approach avoid a much larger configurational entropy work term by beginning with only amino acids, i.e., excluding other organic chemicals and thereby eliminating the "selecting work" which is not accounted for in the 195 kcal/mole calculated above.

In summary, undirected thermal energy is only able to do the chemical and thermal entropy work in polypeptide synthesis, but not the coding (or sequencing) portion of the configurational entropy work. Protenoids are just globs of random polymers. That a polymer

---

*Some studies indicate less than 50% α-links in peptides formed by reacting random mixtures of amino acids. (P.A. Temussi, L. Paolillo, F.E. Benedetti, and S. Andini, 1976. *J. Mol. Evol.* **7**, 105.)

composed exclusively of amino acids (but without exclusively peptide bonds) was formed is a result of the fact that only amino acids were used in the experiment. Thus, the portion of the configurational entropy work that was done—the selecting work—was accomplished not by natural forces but by illegitimate investigator interference. It is difficult to imagine how one could ever couple random thermal energy flow through the system to do the required configurational entropy work of selecting and sequencing. Finally, this approach is of very questionable geological significance, given the many fortuitous events that are required, as others have noted.

Solar Energy

Direct photochemical (UV) polymerization reactions to form polypeptides and polynucleotides have occasionally been discussed in the literature. The idea is to drive forward the otherwise thermodynamically unfavorable polymerization reaction by allowing solar energy to flow through the aqueous system to do the necessary work. It is worth noting that minor yields of small peptides can be expected to form spontaneously, even though the reaction is unfavorable (see eq. 8-16), but that greater yields of larger peptides can be expected only if energy is somehow coupled to the reaction. Fox and Dose have examined the peptide results of Bahadur and Ranganayaki[36] and concluded that UV irradiation did not couple with the reaction. They comment, "The authors do not show that they have done more than accelerate an approach to an unfavorable equilibrium. They may merely have reaffirmed the second law of thermodynamics."[37] Other attempts to form polymers directly under the influence of UV light have not been encouraging because of this lack of coupling. Neither the chemical nor the thermal entropy work, and definitely not any configurational entropy work, has been accomplished using solar energy.

Chemical Energy (Energy-Rich Condensing Agents)

Through the use of condensing agents, the energetically unfavorable dipeptide reaction ($\Delta G_1 = +3000$ cal/mole) is made energetically favorable ($\Delta G_3 < 0$) by coupling it with a second reaction which is sufficiently favorable energetically ($\Delta G_2 < 0$), to offset the energy requirement of the dipeptide reaction:

dipeptide reaction

$$\text{A-OH} + \text{H-B} \longrightarrow \text{A-B} + H_2O \qquad \Delta G_1 > 0 \qquad (9\text{-}5)$$

condensing agent reaction

$$\text{C} + H_2O \longrightarrow \text{D} \qquad \Delta G_2 < 0 \qquad (9\text{-}6)$$

coupled reaction

$$\text{A-OH} + \text{H-B} + \text{C} \longrightarrow \text{A-B} + \text{D} \qquad \Delta G_3 < 0 \qquad (9\text{-}7)$$

As in thermal proteinoid formation, the free water is removed. However, in this case, it is removed by chemical reaction with a suitable condensing agent—one which has a sufficient decrease in Gibbs free energy to drive the reaction forward (i.e., $\Delta G_2 < 0$ and $|\Delta G_2| \geqslant |\Delta G_1|$ so that $\Delta G_1 + \Delta G_2 = \Delta G_3 \leqslant 0$).

Unfortunately, it has proved difficult to find condensing agents for these macromolecule syntheses that could have originated on the primitive earth and functioned properly under mild conditions in an alkaline aqueous environment.[38] Meanwhile, other condensing agents which are not prebiotically significant (e.g., polymetaphosphates) are used in experiments. The plausible cyanide derivative candidates for condensing agents on the early earth hydrolyze readily in aqueous solutions (see Chapter 4). In the process, they do not couple preferentially with the H_2O from the condensation-dehydration reaction. Condensing agents observed in living systems today are produced only by living systems, and thus are not prebiotically significant. Moreover, enzyme activity in living systems first activates amino acids and then brings about condensation of these activated species, thus avoiding the problem of indiscriminate reaction with water.

Notice that if we could solve the very significant problems associated with the prebiotic synthesis of polypeptides by using condensing agents, we would still succeed only in polymerizing random polypeptides. Only the chemical and thermal entropy work would be accomplished by an appropriate coupling of the condensing agent hydrolysis to the condensation reaction. There is no reason to believe that condensing agents could have any effect on the selecting or sequencing of the amino acids. Thus, condensing agents are eliminated as a possible means of doing the configurational entropy work of coding a protein or DNA.

Chemical Energy (Energy-Rich Precursors)

Because the formation of even random polypeptides from amino acids is so energetically unfavorable ($\triangle G = 300$ kcal/mole for 100 amino acids), some investigators have attempted to begin with energy-rich precursors such as HCN and form polypeptides directly, a scheme which is "downhill" energetically, i.e., $\triangle G < 0$. There are advantages to such an approach; namely, there is no chemical work to be done since the bonding energy actually decreases as the energy-rich precursors react to form more complex molecules. This decrease in bonding energy will drive the reaction forward, effectively doing the thermal entropy work as well. The fly in the ointment, however, is that the configurational entropy work is enormous in going from simple molecules (e.g., HCN) directly to complex polymers in a single step (without forming intermediate biomonomers).

The stepwise scheme of experiments is to react gases such as methane, ammonia, and carbon dioxide to form amino acids and other compounds and then to react these to form polymers in a subsequent experiment. In these experiments the very considerable selecting-work component of the configurational entropy work is essentially done by the investigator who separates, purifies, and concentrates the amino acids before attempting to polymerize them. Matthews[39] and co-workers, however, have undertaken experiments where this intermediate step is missing and the investigator has no opportunity to contribute even obliquely to the success of the experiment by assisting in doing the selecting part of the configurational entropy work. In such experiments—undoubtedly more plausible as true prebiotic simulations—the probability of success is, however, further reduced from the already small probabilities previously mentioned. Using HCN as an energy-rich precursor, and ammonia as a catalyst, Matthews and Moser[40] have claimed direct synthesis of a large variety of chemicals under anhydrous conditions. After treating the polymer with water, even peptides are said to be among the products obtained. But as Ferris et al.,[41] have shown, the HCN polymer does not release amino acids upon treatment with proteolytic (protein splitting) enzymes; nor does it give a positive biuret reaction (color test for peptides). In short, it is very hard to reconcile these results with a peptidic structure.

Ferris[42] and Matthews[43] have agreed that direct synthesis of polypeptides has not yet been demonstrated. While some peptide bonds may form directly, it would be quite surprising to find them in significant numbers. Since HCN gives rise to other organic compounds, and various kinds of links are possible, the formation of polypeptides with exclusively α-links is most unlikely. Furthermore, no sequencing would be expected from this reaction, which is driven forward and "guided" only by chemical energy.

While we do not believe Matthews or others will be successful in demonstrating a single step synthesis of polypeptides from HCN, this approach does involve the least investigator interference, and thus, represents a very plausible prebiotic simulation experiment. The approach of Fox and others, which involves reacting gases to form many organic compounds, separating out amino acids, purifying, and finally polymerizing them, is more successful because it involves a greater measure of investigator interference. The selecting portion of the configurational entropy work is being supplied by the scientist. Matthew's lack of demonstrable success in producing polypeptides is a predictable indication of the enormity of the problem of prebiotic synthesis when it is not overcome by illegitimate investigator interference.

Mineral Catalysis

A novel synthesis of polypeptides has been reported[44] which employs mineral catalysis. An aqueous solution of energy-rich aminoacyl adenylates (rather than amino acids) is used in the presence of certain layered clays such as those known as montmorillonites. Large amounts of the energy-rich reactants are adsorbed both on the surface and between the layers of clay. The catalytic effect of the clay may result primarily from the removal of reactants from the solution by adsorption between the layers of clay. This technique has resulted in polypeptides of up to 50 units or more. Although polymerization definitely occurs in these reactions, the energy-rich aminoacyl adenylate (fig. 9-1) is of very doubtful prebiotic significance per the discussion of competing reactions in Chapter 4. Furthermore, the use of clay with free amino acids will not give a successful synthesis of polypeptides. The energy-rich aminoacyl adenylates lower their chemical or bonding energy as they polymerize, driving the reaction forward, and effectively doing the thermal entropy work as well. The role of the clay is to concentrate the reactants and possibly to catalyze the reactions. Once again, we are left with no apparent means to couple the energy flow, in this case in the form of prebiotically questionable energy-rich precursors, to the configurational entropy work of selecting and sequencing required in the formation of specified aperiodic polypeptides, or proteins.

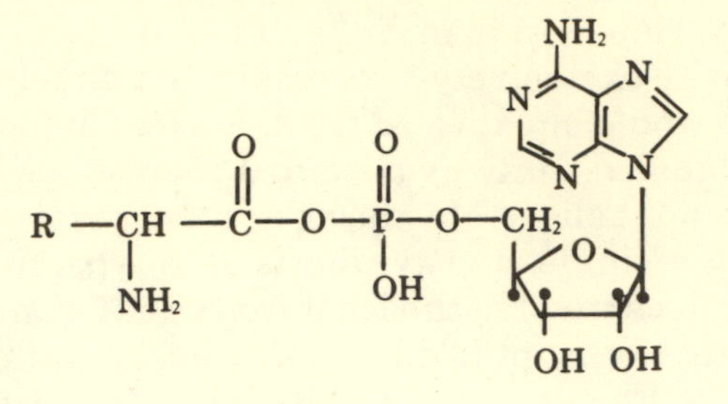

Figure 9-1.
Aminoacyl adenylate.

Summary of Experimental Results on Prebiotic Synthesis of Protein

In summary, we have seen that it is possible to do the thermal entropy work and chemical work necessary to form random polypeptides, e.g., Fox's proteinoids. In no case, though, has anyone been successful in doing the additional configurational entropy work of coding necessary to convert random polypeptides into proteins. Virtually no mechanism with any promise for coupling the random flow of energy through the system to do this very specific work has come to light. The prebiotic plausibility of the successful synthesis of polypeptides must be questioned because of the considerable configurational entropy work of selecting done by the investigator prior to the polymer synthesis. Surely no suggestion is forthcoming that the right composition of just the subset of amino acids found in living things was "selected" by natural means, or that this subset consists only of L-α-amino acids. This is precisely why a large measure of the credit in forming proteinoids must go to Fox and others rather than nature.

Summary of Experimental Results on Prebiotic Synthesis of DNA

The prebiotic synthesis of DNA has proved to be even more difficult than that of protein. The problems that beset protein synthesis apply with greater force to DNA synthesis. Energy flow through the system may cause the nucleotides to chemically react and form a polymer chain, but it is very difficult to get them to attach themselves together in a specified way. For example, 3′-5′ links on the sugar are necessary for the DNA to form a helical structure (see fig. 9-2). Yet 2′-5′ links predominate in most prebiotic simulation experi-

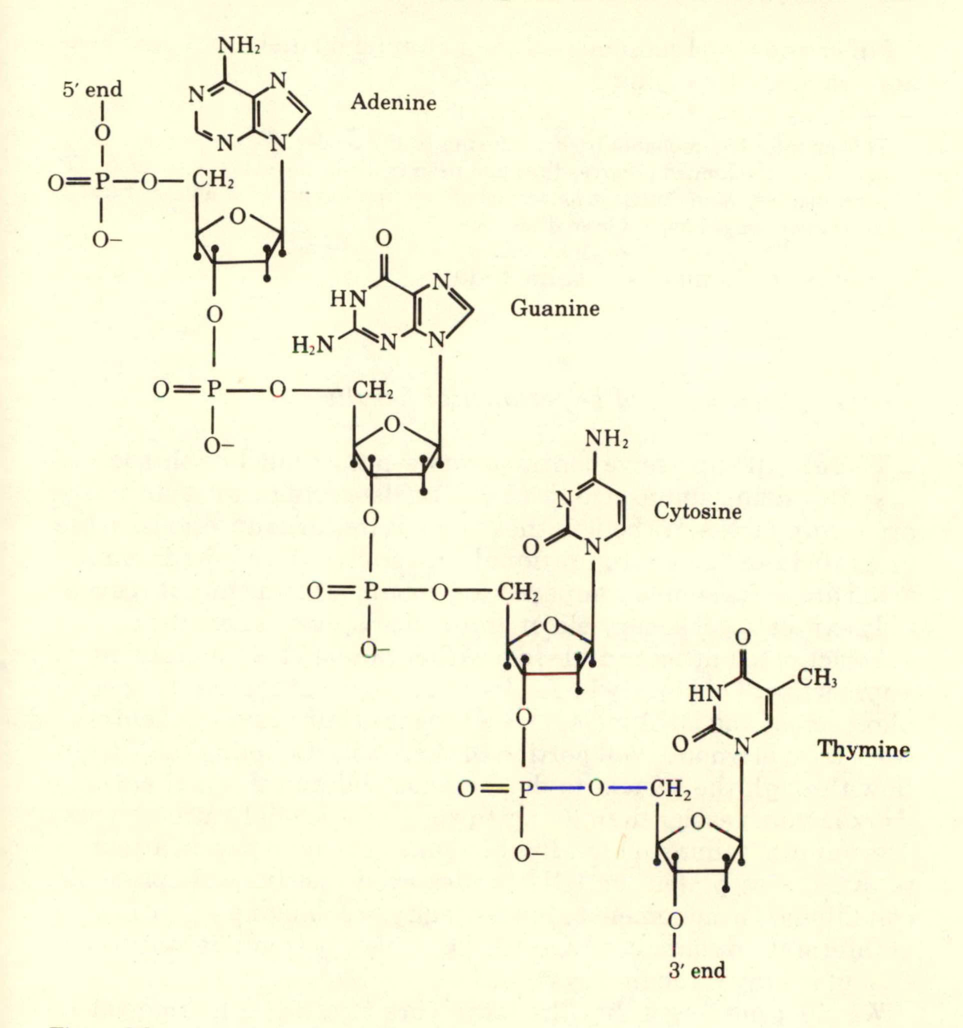

Figure 9-2.
A section from a DNA chain showing the sequence AGCT.

ments.[45] The sequencing of the bases in DNA is also crucial, as is the amino acid sequence in proteins. Both of these requirements are problems in doing the configurational entropy work. It is one thing to get molecules to chemically react; it is quite another to get them to link up in the right arrangement. To date, researchers have only succeeded in making oligonucleotides, or relatively short chains of nucleotides, with neither consistent 3′-5′ links nor specific base sequencing.

Miller and Orgel summarized their chapter on prebiotic condensation reactions by saying:

> This chapter has probably been confusing to the reader. We believe that is because of the limited progress that has been made in the study of prebiotic condensation. Many interesting scraps of information are available, but no correct pathways have yet been discovered.[46]

The situation is much the same today.

Summary Discussion of Experimental Results

There is an impressive contrast between the considerable success in synthesizing amino acids and the consistent failure to synthesize protein and DNA. We believe the reason is the large difference in the magnitude of the configurational entropy work required. Amino acids are quite simple compared to protein, and one might reasonably expect to get some yield of amino acids, even where the chemical reactions that occur do so in a rather random fashion. The same approach will obviously be far less successful in reproducing complex protein and DNA molecules where the configurational entropy work term is a nontrivial portion of the whole. Coupling the energy flow through the system to do the chemical and thermal entropy work is much easier than doing the configurational entropy work. The uniform failure in literally thousands of experimental attempts to synthesize protein or DNA under even questionable prebiotic conditions is a monument to the difficulty in achieving a high degree of information content, or specified complexity from the *undirected* flow of energy through a system.

We must not forget that the total work to create a living system goes far beyond the work to create DNA and protein discussed in this chapter. As we stated before, a minimum of 20-40 proteins as well as DNA and RNA are required to make even a simple replicating system. The lack of known energy-coupling means to do the configurational entropy work required to make DNA and protein is many times more crucial in making a living system. As a result, appeals to chance for this most difficult problem still appear in the literature in spite of the fact that calculations give staggeringly low probabilities, even on the scale of 5 billion years. Either the work—especially the organizational work—was coupled to the flow of energy in some way not yet understood, or else it truly was a miracle.

Summary of Thermodynamics Discussion

Throughout Chapters 7-9 we have analyzed the problems of complexity and the origin of life from a thermodynamic point of view. Our reason for doing this is the common notion in the scientific literature today on the origin of life that an open system with energy and mass flow is *a priori* a sufficient explanation for the complexity of life. We have examined the validity of such an open and constrained system. We found it to be a reasonable explanation for doing the chemical and thermal entropy work, but clearly inadequate to account for the configurational entropy work of coding (not to mention the sorting and selecting work). We have noted the need for some sort of coupling mechanism. Without it, there is no way to convert the negative entropy associated with energy flow into negative entropy associated with configurational entropy and the corresponding information. Is it reasonable to believe such a "hidden" coupling mechanism will be found in the future that can play this crucial role of a template, metabolic motor, etc., directing the flow of energy in such a way as to create new information?

References

1. Albert L. Lehninger, 1970. *Biochemistry*. New York: Worth Publishers, p. 782.
2. H.P. Yockey, 1977. *J. Theoret. Biol.* **67**, 377; R.W. Kaplan, 1974. *Rad. Environ. Biophys.* **10**, 31.
3. M. Eigen, 1971. *Die Naturwiss.* **58**, 465.
4. G. Steinman, 1967. *Arch. Biochem. Biophys.* **121**, 533.
5. A.G. Cairns-Smith, 1971. *The Life Puzzle*. Edinburgh: Oliver and Boyd.
6. F. Crick, 1966. *Of Molecules and Men*. Seattle: University of Washington Press, p. 6-7.
7. Eigen, *Die Naturwiss.*, p. 465; S.L. Miller and L.E. Orgel, 1974. *The Origins of Life on the Earth*. Englewood Cliffs, New Jersey: Prentice-Hall.
8. J.B.S. Haldane, 1965. In *The Origins of Prebiological Systems and of Their Molecular Matrices*, ed. S.W. Fox. New York: Academic Press, p. 11.
9. T. Dobzhansky, 1965. In *The Origins of Prebiological Systems and of Their Molecular Matrices*, p. 310.
10. Ludwig von Bertalanffy, 1967. *Robots, Men and Minds*. New York: George Braziller, p. 82.
11. G. Steinman and M. Cole, 1967. *Proc. Nat. Acad. Sci. U.S.* **58**, 735; Steinman, *Arch. Biochem. Biophys.*, p. 533.

12. A. Katchalsky, 1973. *Die Naturwiss.* **60**, 215; M. Calvin, 1975. *Amer. Sci.* **63**, 169; C.E. Folsome, 1979. *The Origin of Life*. San Francisco: W.H. Freeman, p. 104; K. Dose, 1983. *Naturwiss.* **70**, 378.
13. Steinman, *Arch. Biochem. Biophys.* **121**, 533; Steinman and Cole, *Proc. Nat. Acad. Sci. U.S.*, p. 735.
14. H.P. Yockey, 1981. *J. Theoret. Biol.* **91**, 13.
15. Katchalsky, *Die Naturwiss.*, p. 215.
16. H.R. Hulett, 1969. *J. Theoret. Biol.* **24**, 56.
17. Katchalsky, *Die Naturwiss.*, p. 216.
18. A.E. Wilder-Smith, 1970. *The Creation of Life*. Wheaton, Ill.: Harold Shaw, p. 67.
19. G. Nicolis and I. Prigogine, 1977. *Self Organization in Nonequilibrium Systems*. New York: Wiley.
20. I. Prigogine, G. Nicolis, and A. Babloyantz, 1972. *Physics Today* , p. 23-31.
21. Eigen, *Die Naturwiss.*, p. 465.
22. Prigogine, Nicolis, and Babloyantz, *Physics Today*, p. 23-31.
23. Ibid; Nicolis and Prigogine, *Self Organization in Nonequilibrium Systems*.
24. J.C. Walton, 1977. *Origins,* **4**, 16.
25. P.T. Mora, 1965. In *The Origins of Prebiological Systems and of Their Molecular Matrices*, p. 39.
26. E.R. Harrison, 1969. In *Hierarchical Structures*. ed. L.L. Whyte, A.G. Wilson, and D. Wilson, New York: Elsevier, p. 87.
27. Nicolis and Prigogine, *Self Organization in Nonequilibrium Systems*.
28. Eigen, *Die Naturwiss.*, p. 465; 1971. *Quart. Rev. Biophys.* **4**, 149.
29. Kaplan, *Rad. Environ. Biophysics*, p. 31.
30. J. Brooks and G. Shaw, 1973. *Origin and Development of Living Systems*. New York: Academic Press, p. 209.
31. S.W. Fox and K. Dose, 1977. *Molecular Evolution and the Origin of Life*. New York: Marcel Dekker.
32. P.A. Temussi, L. Paolillo, L. Ferrera, L. Benedetti, and S. Andini, 1976. *J. Mol. Evol.* **7**, 105.
33. S.L. Miller and L.E. Orgel, 1974. *The Origins of Life on Earth*. Englewood Cliffs, New Jersey: Fn. p. 144.
34. C.E. Folsome, 1979. *The Origin of Life*. San Francisco: W.H. Freeman, p. 87.
35. Temussi, Paolillo, Ferrera, Benedetti, and Andini, *J. Mol. Evol.*, p. 105.
36. K. Bahadur and S. Ranganayaki, 1958. *Proc. Nat. Acad. Sci. (India)* **27A**, 292.
37. S.W. Fox and K. Dose, 1972. *Molecular Evolution and the Origin of Life*. San Francisco: W.H. Freeman, p. 142.
38. J. Hulshof and C. Ponnamperuma, 1976. *Origins of Life* **7**, 197.
39. C.N. Matthews and R.E. Moser, 1966. *Proc. Nat. Acad. Sci. U.S.* **56**, 1087; C.N. Matthews, 1975. *Origins of Life* **6**, 155; C. Matthews, J. Nelson, P. Varma, and R. Minard, 1977. *Science* **198**, 622; C.N. Matthews, 1982. *Origins of Life* **12**, 281.
40. C.N. Matthews, and R.E. Moser, 1967. *Nature* **215**, 1230.
41. J.P. Ferris, D.B. Donner, and A.P. Lobo, 1973. *J. Mol. Biol.* **74**, 499.
42. J.P. Ferris, 1979. *Science* **203**, 1135.
43. C.N. Matthews, 1979. *Science* **203**, 1136.
44. Katchalsky, *Die Naturwiss.*, p. 215.
45. R.E. Dickerson, September 1978. *Sci. Amer.*, p. 70.
46. Miller and Orgel, *The Origins of Life on the Earth*, p. 148.

CHAPTER 10

Protocells

A summary of the overall theory of biochemical evolution was given in Chapter 2. Stage 4 of biochemical evolution is the development of protocells, presented in figure 2-1. Protocells represent the link between the synthesis of macromolecules and the appearance of the first living cells. That is, they bridge the gap between the nonliving and the living. It is usually agreed in evolutionary theory that the bridge over this gap is the least understood aspect of the origin of life. William Day has summarized the bridging in the following way:

> In some manner the macromolecules that had condensed from the building blocks managed to associate and pass over the threshold to become life. They assembled into a coordinated arrangement that looked like and functioned as a cell. This was a quantum jump in the events leading to the formation of life and has, of course, because of its spectacular feature, received particular attention.[1]

Types of Protocells

The great chasm in our knowledge of the molecule-to-cell transition means we are free to speculate in many directions. It is not surprising then to see a wide variety of candidates for protocell systems. Some of these are:

1. microspheres (Fox and Dose[2]),
2. coacervates (Oparin[3]),
3. "jeewanu" (Bahadur[4]),
4. NH_4CN microspherules (Labadie et al.[5]),
5. "sulphobes" (Herrera[6]) or "plasmogeny" (Herrera[7]),
6. NH_4SCN-HCHO microstructures (Smith et al.[8]),
7. organic microstructures (Folsome et al.[9]),
8. melanoidin and aldocyanoin microspheres (Kenyon and Nissenbaum[10]), and
9. lipid vesicles (Deamer and Oro,[11] Stillwell[12]).

In 1976, Kenyon and Nissenbaum[13] listed the protocells known at that time (numbers 1-7) and then commented:

> Although each of the proposed model systems exhibits some rudimentary properties of chemical evolutionary interest, it must be emphasized that a very large gap separates the most complex model systems from the simplest contemporary living cells. Moreover, the geochemical plausibility of many of these "protocell" models is open to serious question.[14]

Geochemical Plausibility

Kenyon and Nissenbaum's comment is especially appropriate in view of the evidence cited in the previous chapters. In Chapter 4 we saw that the essential precursor chemicals would probably have been vastly diminished in their concentrations. This conclusion is particularly relevant to the production of protocells, for in all the nine systems proposed above, the organic chemicals must exist in fairly concentrated solutions. That is, the protocell systems proposed are essentially encapsulating mechanisms, and therefore substantial quantities of macromolecules must have existed in close proximity to be enclosed in some primitive membrane. The existence of sufficient concentrations is doubtful, and the lack of geological evidence for a chemical soup or organic ponds supports this pessimistic picture.

The use of high concentrations of selected organic chemicals in the laboratory production of protocells versus the greatly diminished concentrations expected in the ancient geological setting prompted Kenyon and Nissenbaum to comment that "...the geochemical plausibility of many of these 'protocell' models is open to serious question."[15] Several examples will illustrate the implausibility concerning concentrations necessary to form protocells.

Folsome[16] points out that Fox used 15 grams total weight of amino acids in 375 ml of artificial seawater to produce proteinoid microspheres. Therefore, the amino acid concentration would be approximately 0.4 M. Calculations regarding formation rates, concentration rates, and thermal and photochemical decomposition rates point to an abundance of amino acids in seawater of no more than about 10^{-7}M (see Chapter 4). Thus Fox's synthesis uses a molar ratio of amino acids to salts that is "10 million times less in the geologically plausible world."[17]

In more recent experiments, Fox has used concentrations of 6.0 mg of proteinoid per ml of reaction solution.[18] This synthesis would result in proteinoid concentrations of approximately 10^{-3} M, which corresponds to amino acid concentrations of approximately 0.05 M, a figure that is still more than ten thousand times too high to be plausible.

Deamer and Oro state that vesicles of single chain amphiphiles "...require relatively high concentrations [in the millimolar range] of substrate in order to be formed."[19] According to Day, "coacervation can take place in extremely dilute solutions—in concentrations as low as 0.001 percent..."[20] As coacervates are usually formed from relatively high molecular weight compounds (i.e., gum arabic and histone) the molar concentration is also extremely low. The corresponding concentration of the component amino acids would be approximately 10^{-4} M for a 0.001 percent solution. According to Folsome, however, "To make coacervates in the laboratory requires quite high concentrations of polymers."[21] That is, when compared to the primeval ponds of "dilute soup of small organic molecules," Folsome says that a "concentration gap" must be crossed to arrive at the concentration of polymers necessary for coacervation to occur.[22]

The concentrations of amino acids discussed above are typical for the various proposed protocell models. Although the range in concentrations is extremely wide (from 1 to 10^{-4} M), all organic molecules must exist in fairly concentrated solutions relative to geologically plausible concentrations.

In light of the necessary requirements and the conclusions of the previous chapters, it is difficult to imagine that all the correct chemicals or circumstances to form protocells existed on the early earth. Even if the chemicals did occur, large quantities of configurational entropy work would have to be supplied to form biopolymers and then to organize these into a functional cell. As shown in Chapters 8 and 9, unless some hitherto unknown principle operated the availability of such work would have been negligible.

Groups of Protocells

Historically, the two best-known protocell models are the coacervates of Oparin and the proteinoid microspheres of Fox. Lately, Folsome's microstructures and Stillwell's lipid vesicles have also received considerable attention. These models will, therefore, be discussed in more detail.

Stillwell[23] has recently divided the types of protocell models into three groups:

1. Inorganic spheres (Herrera,[24] 1942; Smith et al.,[25] 1968; Grossenbacher and Knight,[26] 1965).
2. Phase-separated polyanions and cations, e.g. Jeewanu (Bahadur,[27] 1972; Bahadur,[28] 1973), coacervates (Oparin,[29] 1968), proteinoid microspheres (Fox and Dose,[30] 1972), and most recently, melanoidin (Kenyon and Nissenbaum,[31] 1976).
3. Lipid vesicles (Goldacre,[32] 1958; Hargreaves and Deamer,[33] 1978).

Stillwell's classification emphasizes the similarity of many of the proposed protocell models. Therefore, although we will not discuss all the models in detail, comments concerning one particular model will typically apply to the whole group. Stillwell's groupings are also relevant, as the following discussions emphasize the actual formation mechanisms of coacervates, microspheres, lipid vesicles, and organic microstructures. By understanding the actual formation processes, the protocell models can be more thoroughly evaluated and the relation among and within groups perceived. The following discussions will focus on groups 2 and 3, as the vast majority of experimental research has been performed on these types of protocell systems.

Coacervates

Coacervates were first noticed by H.G. Bungenberg de Jong in 1932.[34] When nucleic acids, proteins, and other molecules are put into water under certain conditions, spherical droplets 2-670 microns in diameter form. These droplets have higher concentrations of proteins and nucleic acids (compared to the water) and are called coacervates. Oparin realized that coacervates were a potential method to get proteins and nucleic acids together in a concentrated form.

Proteins and nucleic acids have both hydrophilic and hydrophobic parts. Proteins and nucleic acids can also be positively or negatively charged in solution (the charge depending on the pH). The proteins are attracted to the water as are the nucleic acids. If ions (of Na^+, Cl^-, etc.) are added to the solution, they also attract water to themselves. This attraction of water to the ions is usually stronger than the proteins' attraction to the water. Therefore, the water is stripped from the proteins and nucleic acids, making them less soluble. The opposite charges of the nucleic acids and proteins plus the lateral cohesion forces attract the nucleic acids and proteins together to form coacervates. This is why Stillwell groups coacervates as "phase separated" polyanions and cations.[35] This process is sometimes called "salting out" because a salt (Na^+, Cl^-, etc.) is added. The process is based on physical, attractive, and repulsive forces.

Some of the similarities between coacervates and cells noted by Oparin[36] and others are their tendency to form spherical structures, their boundaries, and their ability to absorb selectively. Coacervates are not self-organizing units, however, and they do not contain the structural regularities or selective metabolic processes found in living cells. No matter how large a list of cell-like properties is amassed, the coacervates are simply the result of physical forces of attraction,* and their resemblence to complex living cells is only superficial. We must note, too, that coacervates are formed under very

*By physical forces of attraction, we are referring to the weak interactive forces listed below:

1. Hydrogen bonding.
2. One dipole attracts another dipole (dipole-dipole forces are weak electrostatic attractions).
3. Lateral forces of cohesion (weak forces between like molecules which probably consist of Van der Waals forces).
4. Hydrophilicity and hydrophobicity (result of previous three forces).

These physical forces of attraction are contrasted with forces that form chemical bonds:

1. Sharing of electrons (covalent bonding).
2. Transfer of electrons (ionic bonds).
3. Metallic bonding.

Chemical forces are much stronger than physical ones and usually require a chemical reaction to break and form new bonds. The physical forces, however, do not require any chemical reaction to take place for them to form.

defined conditions of pH, temperature, and ionic strength. They are readily dissolved with dilution, pH change, or heat, and are easily broken up by agitation. In fact, this instability is cited by Fox et al.,[37] and Fox and Dose[38] as evidence that coacervates could have played no major role as intermediate protocells. Coacervates probably would not have existed any length of time in the primitive environment.

Wilder Smith, in his evaluation of coacervates' ability to absorb molecules and increase their mass, states:

> The vital point for us in this whole matter is whether, by means of coacervate formation, we have found any parallel or even insight into biological cell formation, or into the mechanism by which cells increase their mass. That is, whether coacervate formation gives us insight into abiogenesis or into cell metabolism resulting in growth. It is our view that there is absolutely no parallel in the formation of coacervates and protocells. We risk this rather categorical statement on the grounds that there is no evidence that salting-out processes could ever produce anything resembling the inner structure of the true biological cell. For the true biological cell is always, in our experience, so structured and complex that it may be classed as almost one large code in its sequences and specificity. On theoretical grounds alone we do not see any possibility of such structures arising by mere salting-out mechanisms.[39]

He goes on to conclude that:

> It is obvious that coacervate mass increase does not occur by metabolic processes but by purely physical absorption.... In reality, any fundamental likenesses between even the simplest living cells and coacervates are conspicuous by their absence.[40]

The above discussion by Wilder Smith focused on the ability of coacervates to absorb molecules and increase their mass (growth). In the following section, many cell-like properties attributed to proteinoid microspheres will be examined in detail. As shall be observed, many of Wilder Smith's comments above could also apply to proteinoid microspheres.

Microspheres

Microspheres form when solutions of proteinoids cool. A "remarkable" list of cell-like properties has been assembled by Fox and Dose,[41] and Fox et al.,[42] (provided in table 10-1) and most recently by Fox and Nakashima.[43] Some microspheres are shown in figure 10-1.

Table 10-1.
Proteinoid microparticles possess many properties similar to contemporary cells.

Stability (to standing, centrifugation, sectioning)
Microscopic size
Variability in shape but uniform in size
Numerousness
Stainability
Producibility as gram-positive or gram-negative
Osmotic type of property in atonic solutions
Ultrastructure (electron microscope)
Double-layered boundary
Selective passage of molecules through boundary
Catalytic activities
Patterns of association
Propagation by "budding" and fission
Growth by accretion
Motility
Selective inclusion of polynucleotides with basic proteinoids (particles are composed of nucleoproteinoid not proteinoid)

(From S.W. Fox, K. Harada, G. Krampitz, and G. Mueller, June 22, 1970. *Chem. Eng. News*, p. 90.)

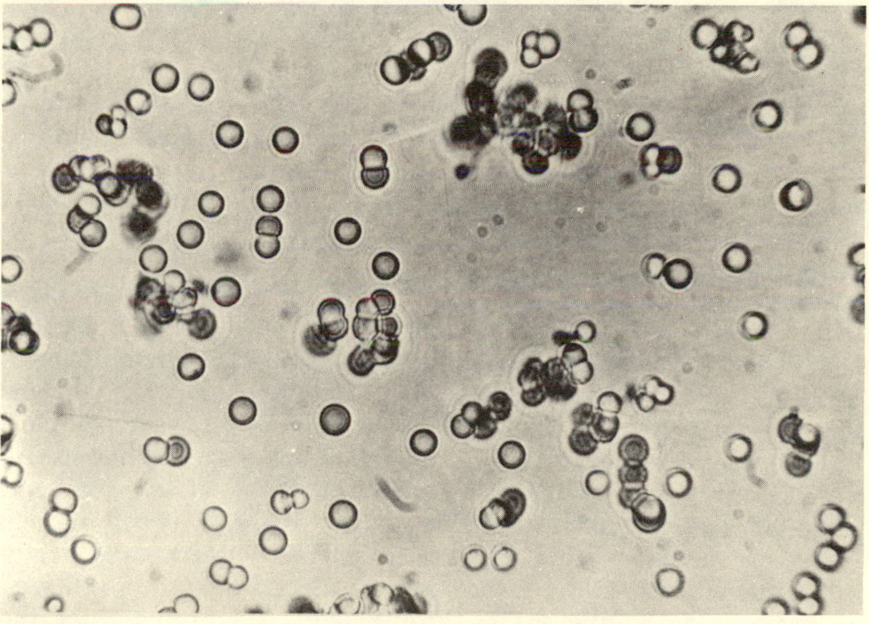

Figure 10-1.
Optical micrograph of protenoid microspheres. Microspheres formed when an amino acid polymer (proteinoid) was boiled in water. The proteinoid resulted from pyrocondensation of dry amino acids. (From S.W. Fox and K. Dose, *Molecular Evolution and the Origin of Life*, revised ed., p. 214.)

Because of the many similar properties between microspheres and contemporary cells, microspheres were confidently called protocells, the link between the living and nonliving in evolution. Similar structures were given the names plasmogeny[44] (plasma of life) and Jeewanu[45] (Sanskrit for "particles of life").

Essentially, microspheres result when small "protein-like" substances (proteinoids) are placed in water. As previously stated, proteinoids have both hydrophilic and hydrophobic parts. When the concentration of the proteinoids is increased, the lateral forces of cohesion between the proteinoids bring them together into a spherical particle (technically called an association colloid). These particles can also form micelles, structural aggregates in which the hydrophilic part of the protein extends outward into the water and the hydrophobic part inward.

Kenyon and Steinman also emphasize the role of micelles:

> Large molecules with both polar and nonpolar regions have the ability to form micelles in aqueous solutions. This phenomenon results from the nonpolar regions of several such molecules coming close enough together to mutually exclude much of the water in their immediate vicinity. At the same time the polar ends face outwards to the aqueous environment.[46]

Likewise, microspheres are simply proteinoids attracted together (by physical forces) into a somewhat ordered spherical structure. Here too, the structure is due to the attraction of the hydrophilic parts of the proteinoids to water and of the hydrophobic parts to each other.

We will examine in detail some of the "cell-like" properties of microspheres. Fox et al., state that "microparticles possess in large degree the rate enhancing activities of the polymer of which they are composed."[47] These are microspheres' "catalytic activities" listed in table 10-1. If the protein by itself has a catalytic property, it seems very logical that the protein would retain that property when put in a micelle. The catalytic activity of the microsphere is not due to any special structure that the microsphere possesses. The increase in reaction rate observed in microspheres is very small by comparison to the rate increase seen in true enzymes (where rate increase factors are in the billions—10^9). Furthermore, much of the rate increase seen in proteinoids is due to the amino acids themselves, not the proteinoid.

Another "cell-like" property cited is the selective passage of certain molecules. Fox et al., explain that "Polymers that are similar in composition to those inside the microspheres can selectively diffuse through the boundary."[48] It is to be expected that similar molecules (the hydrophobic ones) would be incorporated into the micelle by the physical forces of attraction present.

Microspheres also "grow by accretion"[49] (see figure 10-1). This, however, is the attraction of like molecules to the micelle by simple physical forces. The process of microsphere "growth" has little if any similarity to the process by which contemporary cells grow. True cells grow through a metabolic process involving many chemical reactions. In microspheres no chemical reactions are taking place, only accumulation through physical forces of attraction.

"Propagation by budding"[50] (see figure 10-1) also has no connection to the present day cell process of reproduction, which requires enzymes, DNA, energy, and many reactions coupled together precisely. By contrast, the "budding" illustrated in microspheres is merely a breaking up of the microsphere due to heat or pH changes.

Oparin further criticizes microspheres by saying,

> Fox's microspheres, since they are obtained thermally, do not present very promising results from this view [i.e., evolving to include metabolic processes]. Their structure is static. This...creates difficulties when it comes to converting them into dynamic systems which could be used for modeling the evolution of metabolism.[51]

Miller and Orgel also criticize Fox's statements relating microspheres to living cells. They state that the microsphere's bilayer membranes "...are not 'biological-like' membranes since they do not contain lipids or carry out any of the functions of biological membranes."[52] They conclude, "It seems unlikely...that the division of microspheres is related to the origin of cell division."[53]

One of the most important aspects of any cell is its chemical composition. As mentioned in Chapter 9, proteinoids (from which microspheres are formed) contain many nonbiological features. In fact, Temussi refers to proteinoids as "the preferential formation of unnatural peptide bonds."[54]

Folsome criticizes microspheres in that they possess a "grossly thick" boundary layer that more closely resembles a nearly impermeable cell wall or spore coat than a cell membrane.[55]

In the present-day cell, there are thousands of different chemical reactions taking place. Not even one chemical reaction takes place in microspheres, only mechanical and physical processes due to simple attractive forces. We question listing these purely physical forces as resemblances to true cell processes. In truth, they have scant relation to actual processes in living cells. Actually, microspheres possess only outward likenesses and nothing of the inward structure and function of a true cell. They contain no information content, no energy utilizing system, no enzymes, no nucleic acid, no genetic code, and no replication system. They contain only a mixture of polymers of amino acids, the so-called proteinoids. Microspheres cannot be said to be living in any sense of the word, and it is questionable whether they should even be given the name "protocell." They are merely an aggregation of polymers, and do not help to bridge the gap between life and non-life.

Also mentioned previously are the unlikely geological conditions that would be necessary to form microspheres. The requirement of implausible conditions has been emphasized by Miller and Urey[56] and Miller and Orgel.[57] In reference to Fox's method of microsphere preparation, Folsome asks, "The central question is where did all these pure, dry, concentrated, and optically active amino acids come from in the real, abiological world?"[58]

William Day reflects similar views concerning microspheres, coacervates, and Jeewanu when he states, "There have been similar efforts to create models of the primal cell where a greater regard was given to the gross morphology than chemical functionality."[59] But, says Day, "No matter how you look at it, this is scientific nonsense."[60] Finally, Day concludes:

> These pseudo-cellular models, like clay, soap bubbles, or any other inanimate objects, have neither the mechanism nor the potential of becoming anything beyond what they are.... But the most serious fault of models from particles held together by ionic forces is that they would have been continually periled with dissolution. Coacervates are notoriously unstable and microspheres exist only in saturated solutions. Their existence in Archean lakes or oceans would have been short-lived.[61]

In his critique of microspheres and coacervates, Folsome emphasizes that these models "...suffer from the same practical problems of the concentration gap."[62] That is, the formation of microspheres and coacervates requires quite high concentrations of polymers not present in the primeval ponds. Folsome goes on to say, "Hypothetically, there are ways to circumvent the concentration gap, but all appear to be more wishful thinking than plausible facets of reality."[63]

Overall, it appears that coacervates, microspheres, and all the "phase-separated polyanion and cation" models of group 2 have serious deficiencies that disqualify them as protocell systems. That is, they cannot be considered forerunners to the modern cell.

Lipid Vesicles

The interest in lipids stems from their functionality in modern membranes. Here they have a primary role, and it is not surprising they should be used in developing protocell systems. Bangham and Horne[64] originally demonstrated that phospholipid molecules will self-assemble into closed vesicles. Phospholipids are fatty acid derivatives of glycero-phosphoric acid. The hydrocarbon chain of the fatty acid is hydrophobic, whereas the phosphate end of the molecule is hydrophilic. Therefore, the phospholipids align themselves when surrounded by water to form spherical shapes. If a single layer of phospholipid molecules forms, a micelle results. If a bimolecular layer creates a sphere, the particle is a liposome or vesicle.

Simple fatty acids with hydrocarbon chains of eight or more carbons can also form structured vesicles or micelles, depending on the pH of the solution.[65] Compared to liposomes, however, the structures are relatively unstable, and quite sensitive to ionic environment and temperature. They also require relatively high concentrations (in the millimolar range) in order to form.[66] Contemporary phospholipids can form vesicles at lower concentration, and are not so sensitive to the environment. Such vesicles have been criticized as being composed of highly evolved phospholipids.[67] Stillwell[68] states that phospholipids were probably not present in the early ocean, while Deamer and Oro[69] claim that phospholipids can be formed under plausible prebiotic conditions. In the opinion of the authors, however, several nongeological (i.e., implausible) chemical components have been used in the synthesis. In particular, soluble phosphate compounds were used as reactants. It is doubtful, however, that soluble phosphate concentration exceeded 10^{-6} M in the primitive ocean, due to precipitation by calcium and magnesium salts. The fatty acids needed for phospholipid formation would also predictably have been in short supply in the oceanic soup, having precipitated with calcium and magnesium salts. (See Chapter 4)

The synthesis of complex lipids, such as the phospholipids, probably also suffered from the concentration gap discussed earlier. The precursors of the complex lipids include fatty acids, glycerol, and glycerol phosphate.[70] These compounds, if they existed at all in the prebiotic soup, would have been present in dilute concentrations, since they would have been subject to many competing reactions. In view of this, the formation of more complex lipids necessary for stable vesicles is dubious.

Note that in Stillwell's[71] review of lipid membranes in protocells, he criticizes microspheres and coacervates as being too "leaky" to be protocells. That is, the molecules encapsulated in the structure can easily leak through the boundary. Interestingly, the lipid vesicles may be too "tight." They do not readily transport molecules through their membrane. Contemporary cells contain both lipids and proteins in their membranes, enabling a complex selective transport mechanism to operate. Several transport mechanisms have been proposed for the vesicle protocells.[72] The facilitated diffusion of molecules through the boundary may be one of the few function-like properties of vesicles. Nevertheless, the mechanisms are nowhere comparable to those in contemporary cells. In summary, the vesicle protocells bear only superficial resemblance to true cells.

Organic Microstructures

Folsome[73] has been the main proponent of the organic microstructure protocell system. Microstructures are formed during Miller-Urey electrical discharge experiments. They resemble (morphologically) microfossils found in ancient rocks and are thought to consist of cross-linked kerogenous polymer structures.

Fox has criticized Folsome's experiments as being nongeological—"without terrestrial counterpart."[74] Furthermore, Fox states that the alleged potential of microstructures as a protocell model is poorly supported because the microstructures have not demonstrated any cellular function. The microstructures are also basically uncharacterized and not shown to contain polymers.

Fox's criticisms appear valid. A strength of Folsome's structures which must be acknowledged, however, is that they do not require the usual stepwise approach. That is, the organic microstructures form directly in spark discharge simulation experiments. This is in contrast to the formation of most protocell models, which require intermediate steps. For example, coacervates are formed from relatively high molecular weight polymers such as histones and gum arabic, and microspheres are formed from pure amino acids. In view of the discussion in Chapter 5, however, concerning the composition of the primitive earth and atmosphere, the geological plausibility of Folsome's highly reducing, closed-flask experiments should be questioned. The limited evidence (first-order kinetics and self-assembly) given by Folsome in support of the organic microstructure's biogenicity suffers from the same problems as other proposed protocell systems. That is, purely physical and morphological properties are being dressed up to resemble present-day cellular processes when no true functional similarity exists. In fact, the morphology of organic microstructures is very diverse and sometimes irregular. Present-day cells are typically spherical with smooth, regular boundaries. Organic microstructures possess few, if any, properties of present-day cells and must therefore be questioned as forerunners of true cells.

Conclusion

In light of the conclusions from the previous chapters (especially Chapters 4, 5, 8, and 9), it seems doubtful that the macromolecules necessary for living cells existed on the early earth. Even if the molecules were present in substantial quantities, the encapsulating protocell systems reviewed in this chapter appear to be highly tenuous as true protocells. In most cases, the only resemblance that the proposed models have to contemporary cells is their size and morphology (spherical shapes).

Cellular functions claimed for the protocell system are the result of simple physical forces. Similarities to present-day cell processes are superficial. In all cases, the protocell systems are only conglomerations of organic molecules that provide no genuine steps to bridge the gap between living and nonliving. Furthermore, most protocells are highly unstable and were formed under nongeological conditions. In summary, the assessment of Green and Goldberger is still appropriate:

> ...the macromolecule-to-cell transition is a jump of fantastic dimensions.... The available facts do not provide a basis for postulating that cells arose on this planet.[75]

References

1. William Day, 1979. *Genesis on Planet Earth*. East Lansing, Mich.: The House of Talos Publications, p. 310.
2. Sidney W. Fox and Klaus Dose, 1972. *Molecular Evolution and the Origin of Life*. San Francisco: W.H. Freeman, p. 198.
3. A.I. Oparin, 1959. In *The Origin of Life on the Earth*, ed. A.I. Oparin, p. 301-321; 1965. In *The Origin of Prebiological Systems*, ed. S.W. Fox. New York: Academic Press, p. 331.

4. K. Bahadur, 1966. *Synthesis of Jeewanu, the Protocell*. Allahabad: Ram Narain Lal Beni Prasad.
5. M. Labadie, G. Cohere, C. Brechenmacher, 1967. *Compt. Rend. Soc. Biol.* **161**, 1689.
6. A.L. Herrera, 1940. *Bull. Lab. Plasmogenie, Mex.* **2**, 3.
7. A.L. Herrera, 1942. *Science* **96**, 14.
8. A.E. Smith, J.J. Silver, and G. Steinman, 1968. *Experientia* **24**, 36.
9. C.E. Folsome, R.D. Allen, N. Ichinose, 1975. *Precambrian Res.* **2**, 263.
10. A. Nissenbaum, D.H. Kenyon, and J. Oro, 1975. *J. Mol. Evol.* **6**, 253; D.H. Kenyon, and A. Nissenbaum, 1976. *J. Mol. Evol.* **7**, 245.
11. D.W. Deamer and J. Oro, 1980. *BioSystems* **12**, 167.
12. W. Stillwell, 1980. *Origins of Life* **10**, 277.
13. Kenyon and Nissenbaum, *J. Mol. Evol.*, p. 246.
14. Ibid., p. 246.
15. Ibid., p. 246.
16. C.E. Folsome, 1977. *Die Naturwiss.* **64**, 381.
17. Ibid., p. 381.
18. S.W. Fox and T. Nakashima, 1980. *BioSystems* **12**, 155.
19. Deamer and Oro, *BioSystems*, p. 171.
20. Day, *Genesis on Planet Earth*, p. 313.
21. C.E. Folsome, 1979. *The Origin of Life, a Warm Little Pond*. San Francisco: W.H. Freeman, p. 83.
22. Ibid., p. 83.
23. Stillwell, *Origins of Life*, p. 277.
24. Herrara, *Science*, p. 14.
25. Smith, Silver, and Steinman, *Experientia*, p. 36.
26. K.A. Grossenbacher and C.A. Knight, 1965. In *The Origin of Prebiological Systems*, p. 173.
27. K. Bahadur, 1972. *Zbl. Bakt.* **127**, 643.
28. K. Bahadur, 1973. *India Nat. Sci. Acad.* **39B**, 455.
29. A.I. Oparin, 1968. *Genesis and Evolutionary Development of Life*. New York: Academic Press.
30. Fox and Dose, *Molecular Evolution and the Origin of Life*, p. 198.
31. Kenyon and Nissenbaum, *J. Mol. Evol.*, p. 245.
32. R.J. Goldacre, 1958. In *Surface Phenomena in Chemistry and Biology*, ed. J.F. Danielli. New York: Pergamon Press, p. 278.
33. W.R. Hargreaves and D.W. Deamer, 1978. In *Light Tranducing Membranes, Structure, Function and Evolution*, ed. D.W. Deamer. New York: Academic Press.
34. H.C. Bungenberg de Jong, 1932. *Protoplasma* **15**, 110; (in Oparin) A.I. Oparin, 1957. *The Origin of Life on the Earth*.
35. Stillwell, *Origins of Life*, p. 277.
36. Oparin, 1957. *The Origin of Life on the Earth*, p. 301.
37. S.W. Fox, K. Harada, G. Krampitz, and G. Mueller, June 22, 1970. *Chem. Eng. News*, p. 80.
38. Fox and Dose, *Molecular Evolution and the Origin of Life* , p. 220.
39. A.E. Wilder Smith, 1970. *The Creation of Life*. Wheaton, Ill.: Harold Shaw Publishers, p. 84-85.
40. Ibid., p. 85.
41. Fox and Dose, *Molecular Evolution and the Origin of Life*, p. 233.
42. Fox et al., *Chem. Eng. News*, p. 90.

43. Fox and Nakashima, *BioSystems*, p. 155.
44. Herrera, *Science*, p. 14.
45. Bahadur, *Synthesis of Jeewanu, the Protocell.*
46. Dean H. Kenyon and Gary Steinman, 1969. *Biochemical Predestination.* New York: McGraw-Hill, p. 251.
47. Fox et al., *Chem. Eng. News*, p. 92.
48. Ibid.
49. Ibid.
50. Ibid.
51. Oparin, *Genesis and Evolutionary Development of Life*, p. 105.
52. Stanley L. Miller and Leslie E. Orgel, 1974. *The Origins of Life on the Earth.* Englewood Cliffs, New Jersey: Prentice-Hall, p. 144.
53. Ibid.
54. P.O. Temussi, 1976. *J. Mol. Evol.* **8**, 305.
55. Folsome, *The Origin of Life*, p. 87.
56. S.L. Miller and H.C. Urey, 1959. *Science* **130**, 245.
57. Miller and Orgel, *The Origins of Life on the Earth*, p. 145.
58. Folsome, *The Origin of Life*, p. 87.
59. Day, *Genesis of the Planet Earth*, p. 319.
60. Ibid., p. 319.
61. Ibid., p. 320.
62. Folsome, *The Origin of Life*, p. 85.
63. Ibid., p. 84.
64. A.D. Bangham and R.W. Horne, 1964. *J. Mol. Evol.* **8**, 660.
65. Deamer and Oro, *BioSystems*, p. 171.
66. Ibid.
67. Stillwell, *Origins of Life*, p. 290.
68. W. Stillwell, 1976. *BioSystems* **8**, 111.
69. Deamer and Oro, *BioSystems*, p. 173.
70. Ibid., p. 171.
71. Stillwell, *BioSystems*, p. 111.
72. Ibid., p. 112.
73. Folsome, Allen, and Ichinose, 1975. *Precambrian Res.*, 263; Folsome, *The Origin of Life*. p. 87-90.
74. S.W. Fox, 1977. *Die Naturweiss.* **64**, 380.
75. D.E. Green and R.F. Goldberger, 1967. *Molecular Insights into the Living Process.* New York: Academic Press, p. 407.

CHAPTER 11

Summary and Conclusion

Summary

Chemical evolution is broadly regarded as a highly plausible scenario for imagining how life on earth might have begun. It has received support from many competent theorists and experimentalists. Ideas of chemical evolution have been modified and refined considerably through their capable efforts. Many of the findings of these workers, however, have not supported the scenario of chemical evolution. In fact, what has emerged over the last three decades, as we have shown in the present critical analysis, is an alternative scenario which is characterized by destruction, and not the synthesis of life.

This alternative scheme envisions a primitive earth with an oxidizing atmosphere. A growing body of evidence supports the view that substantial quantities of molecular oxygen existed very early in earth history before life appeared. If the early atmosphere was strongly oxidizing, as we find on Mars today, then no chemical evolution ever occurred. Even if the primitive atmosphere was reducing or only mildly oxidizing, then degradative processes predominated over synthesis. Furthermore, macromolecule polymerization would be subjected to countless competing reactions. Small steady-state concentrations (no greater than 10^{-7}M for amino acids, for example) of essential precursor chemicals would fill the earth's water basins. Because of such small concentrations the rates of chemical evolution in the ocean were never more than negligible. This follows from the law of mass action. The same law also predicts that any concentrating mechanisms (such as freezing or evaporating ponds) would merely have served to accelerate *both* destructive and synthetic processes already going on at slower rates in the dilute seas. In the end there would have been no discernible chemical evolutionary benefit from these small concentrating ponds. An idea of how dilute in biomonomers these seas must have been comes from the fact that the prebiotic chemical soup, presumably a world-wide phenomenon, left no known trace in the geological record.

Since monomer concentrations were so low, polymerizations by spontaneous means were made all the more difficult. The primary difficulty was not lack of suitable energy sources. Rather it was both a lack of sufficient energy mobilizing means to harness the energy to the specific task of building biopolymers *and* a lack of means to generate the proper sequence of, say, amino acids in a polypeptide to get biological function. We have identified this latter problem as one of doing the configurational entropy work. Here the difficulty is fundamental. It applies equally to discarded, present, and possible future models of chemical evolution. We believe the problem is analagous to that of the medieval alchemist who was commissioned to change copper into gold. Energy flow through a system can do chemical work and produce an otherwise improbable distribution of energy in the system (e.g., a water heater). Thermal entropy, however, seems to be physically independent from the information content of living systems which we have analyzed and called configurational entropy. As was pointed out, Yockey has noted that negative thermodynamic entropy (thermal) has nothing to do with information, and no amount of energy flow through the system and negative thermal entropy generation can produce even a small amount of information. You can't get gold out of copper, apples out of oranges, or information out of negative thermal entropy. There does not seem to be any physical basis for the widespread assumption implicit in the idea that an open system is a sufficient explanation for the complexity of life. As we have previously noted, there is neither a theoretical nor an experimental basis for this hypothesis. There is no hint in our experience of any mechanistic means of supplying the necessary configurational entropy work. Enzymes and human intelligence, however, do it routinely.

Actually the configurational entropy work is of two types. The job of selecting or sorting the appropriate chemical composition out of a random soup mixture we have referred to as the "selecting" work. The task of arranging these selected monomers in the proper sequence in a polymer for biological function is the "coding" work. The early earth conditions appear to offer no intrinsic means of supplying either of these indispensable components of the configurational entropy work necessary to make the macromolecules of life.

It is this unmet configurational entropy work requirement which is the central problem in developing essential macromolecules such as DNA and protein, much less the complex cellular structures.

So-called protocells have been produced in the laboratory in an attempt to bridge the nonliving and the living. Such structures do have the crude resemblance to true cells but none of the internal cellular machinery, such as enzymes, DNA, or phospholipid cell membranes. The few "cell" functions manifested by protocell systems typically arise from simple physical forces. Any similarity to true cellular processes is highly superficial.

The usual interpretation of chemical evolution derives a great deal of apparent plausibility from reports of laboratory prebiotic simulation experiments. In fact most of these experiments are probably invalid. Unlike other established experimental disciplines, "prebiotic chemistry" has no generally accepted criterion for what constitutes a valid prebiotic simulation experiment. Consequently, many incredible experiments have been published as "simulation" experiments.

As a meager step toward remedying this situation, we have offered a tentative definition of a valid prebiotic simulation experiment. Based on the widely held view that life was not the result of the *crucial* involvement of the supernatural, we have carefully extended this to show that a valid prebiotic simulation experiment must not have crucial investigator interference in any *illegitimate* sense. By definition there are numerous *legitimate* activities of the investigator. Simply stated, an investigator may appropriately adjust conditions of the experiment that are deemed analogous to the primitive earth situation. But such conditions must be plausible.

To help evaluate the degree of interference by the investigator, we devised a scale on which we placed the various common experimental procedures. To the degree a lab experiment deviates from plausible early earth conditions, to that degree it is an illegitimate interference by the experimenter. This view assumes that we need to take into account only *probable* conditions. With the help of this scale we have judged conditions of most simulation experiments to be implausible, and therefore excluded them as legitimate simulation experiments. This is a severe judgement. But it should be recognized that part of the deep suspicion that has surrounded prebiotic chemistry from its beginning has been over precisely this matter that ill-defined experimental criteria have been used. As one scoffer was heard to remark in a scientific meeting, "In prebiotic chemistry anything counts." It is up to the investigators in this field to come to grips with the problem of what is a valid simulation experiment, and what is not.

One characteristic feature of the above critique needs to be emphasized. We have not simply picked out a number of details within chemical evolution theory that are weak, or without adequate explanation *for the moment*. For the most part this critique is based on crucial weaknesses intrinsic to the theory itself. Often it is contended that criticism focuses on present ignorance. "Give us more time to solve the problems," is the plea. After all, the pursuit of abiogenesis is young as a scientific enterprise. It will be claimed that many of these problems are mere state-of-the-art gaps. And, surely, some of them are. Notice, however, that the sharp edge of this critique is not what we *do not* know, but what we *do* know. Many facts have come to light in the past three decades of experimental inquiry into life's beginning. With each passing year the criticism has gotten stronger. The advance of science itself is what is challenging the notion that life arose on earth by spontaneous (in a thermodynamic sense) chemical reactions.

Over the years a slowly emerging line or boundary has appeared which shows observationally the limits of what can be expected from matter and energy left to themselves, and what can be accomplished only through what Michael Polanyi has called "a profoundly informative intervention."[1] When it is acknowledged that most so-called prebiotic simulation experiments actually owe their success to the crucial but *illegitimate* role of the investigator, a new and fresh phase of the experimental approach to life's origin can then be entered. Until then however, the literature of chemical evolution will probably continue to be dominated by reports of experiments in which the investigator, like a metabolizing Maxwell Demon, will have performed work on the system through intelligent, exogenous intervention. Such work establishes experimental boundary conditions, and imposes intelligent influence/control over a supposedly "prebiotic" earth. As long as this informative interference of the investigator is ignored, the illusion of prebiotic simulation will be fostered. We would predict that this practice will prove to be a barrier to solving the mystery of life's origin.

Conclusion

A major conclusion to be drawn from this work is that the undirected flow of energy through a primordial atmosphere and ocean is at present a woefully inadequate explanation for the incredible complexity associated with even simple living systems, and is probably wrong.

Many will find this critique "interesting" but will not draw the same conclusions we have. Why will many predictably persist in their acceptance of some version of chemical evolution? Quite simply, because chemical evolution has not been *falsified*. One would be irrational to adhere to a falsified hypothesis. We have only presented a case that chemical evolution is highly implausible. By the nature of the case that is all one *can* do. In a strict, technical sense, chemical evolution *cannot* be falsified because it is not falsifiable. Chemical evolution is a speculative reconstruction of a unique past event, and cannot therefore be tested against recurring nature. As Pirie remarked, "Now we have little expectation of being able to conclude a discussion with the statement 'This is how life did arise'; the best we can hope for is 'This is one of the ways life could have arisen.' "[2]

Some will immediately conclude that if Pirie is right, then chemical evolution is not science and it should be consigned to the rubbish heap. This seems to have been the conclusion of Mora when he said, "...how life originated, I am afraid that, since Pasteur, this question is not within the scientific domain."[3]

But this conclusion is too hasty. It must be realized, as we pointed out in Chapter 1, that the speculative nature of chemical evolution does not mean that it is without value. In forensic medicine, a speculative scenario in the hands of a skillful lawyer can be used to persuade a jury of the guilt or innocence of a defendant. So it is with chemical evolution scenarios.

In the persuading process there is always the risk that partial truth will be viewed as the whole truth and mislead a jury. To minimize the risks of convicting the innocent and freeing the guilty, the court in the U.S.A. uses an adversarial approach, which means the jury gets to hear likely scenarios from attorneys for both prosecution and defense. In addition, attorneys from both sides can cross-examine witnesses. When a jury weighs the evidence, it is hoped the evidence in hand is a fair sampling so that justice is served. For a jury to render a guilty verdict for a capital offense the case must be established beyond reasonable doubt.

To be sure, the case for the origin of life via chemical evolution as usually presented sounds plausible, and has been accepted very widely, if not generally, by the scientific community. Furthermore, popularizations have carried the case to millions in a persuasive manner. Because of the fact that chemical evolution cannot be falsified, however, its apparent plausibility can easily be exaggerated beyond its true status as speculation and be regarded instead as knowledge.

Perhaps this is always a danger with speculative approaches, but it would seem to be particulary likely here since the substantial case questioning the plausibility of chemical evolution has been all but muted. Our chapters, we believe, have shown that reasonable doubt exists concerning whether simple chemicals on a primitive earth did spontaneously evolve (or organize themselves) into the first life. We leave it to the jury to decide.

> "That's the worst of circumstantial evidence. The prosecuting attorney has at his command all the facilities of organized investigation. He uncovers facts. He selects only those which, in his opinion, are significant. Once he's come to the conclusion the defendant is guilty, the only facts he considers significant are those which point to the guilt of the defendant. That's why circumstantial evidence is such a liar. Facts themselves are meaningless. It's only the interpretation we give those facts which counts."
>
> "Perry Mason"—Erle Stanley Gardner*

References

1. Michael Polanyi, 1967. *Chem Eng. News*, Aug. 21, p. 54.
2. N.W. Pirie, 1957.*Annals of the New York Academy of Science* **66**, 369.
3. P.T. Mora, 163. *Nature* **199**, 212.

**The Case of the Perjured Parrot*, Copyright 1939, by Erle Stanley Gardner.

Epilogue

In the introductory chapter we stated our hope that criticism of current theories of the origin of life would prove to be a first step toward a more satisfactory theory of origins. No consideration, however, was given to alternatives. So, in this epilogue we will consider five alternative views which have been mentioned in the literature on the origin of life. These are:

1. New natural laws
2. Panspermia
3. Directed Panspermia
4. Special Creation by a creator within the cosmos
5. Special Creation by a Creator beyond the cosmos

We foresee that the major theories of origins for the future are listed here. Before considering these, however, let us enumerate some notable results from our analysis of origin of life research. Any satisfactory alternative should account for these factors:

1. There is accumulating evidence for an oxidizing early earth and atmosphere.
2. Destructive processes would have predominated over synthesis in the atmosphere and ocean in the prebiotic world.
3. There is continued shortening of the time interval (now $<$ 170 my) between earth's cooling and the first appearance of life.
4. Geochemical analysis shows the composition of Precambrian deposits is short of nitrogen.

5. There is an observational limit or boundary between what has been accomplished in the laboratory by natural processes left to themselves and what is done through investigator interference.
6. In our experience only biotic processes (enzymes, DNA, etc.) and investigator interference couple energy flow to the task of constructing biospecific macromolecules.
7. True living cells are extraordinarily complex, well orchestrated dynamic structures containing enzymes, DNA, phospholipids, carbohydrates, etc., to which so-called protocells bear only a superficial resemblance.

New Natural Laws*

We have seen the failure, perhaps the impotence of presently known fundamental physical and chemical laws to explain the origin of biological structures. This has given renewed inspiration to the idea that new principles of physics must be discovered to adequately explain this phenomenon. Elsasser[1] has argued that classes of living structures are too small to be subject to the statistical averaging procedures of physics, suggesting that new natural laws must be identified instead. Recall from Chapter 1 that this was also the suggestion of Murray Eden at the Wistar Institute Symposium. In the same vein Garstens[2] postulated that the application of statistical mechanics to biological systems requires a new set of auxiliary assumptions different from those traditionally used in physics. Mora[3] concurs that new laws are essential, pointing out that it is impossible to reconcile statistical and thermodynamic constraints with the formation of living systems.

Using the quantum mechanical method, Wigner[4] calculated the probability of a living organism interacting with nutrients to produce another identical organism, assuming that this interaction is governed by a random symmetric Hamiltonian matrix. This is the same assumption employed by von Neumann[5] to prove that the Second Law of Thermodynamics is a consequence of quantum mechanics. On counting up the number of equations describing the interactions, Wigner found they greatly exceeded the number of unknowns which described the final state of the nutrient plus two

*This section draws heavily on the theoretical analysis of John C. Walton in an article entitled "Organization and the origin of life," in *Origins*, vol. 4, no. 1, 1977, 16-35.

organisms. Wigner's analysis showed a zero probability that there would be any state of the nutrient which would allow multiplication of the organism. He says: "It would be a miracle" and would imply the interaction of the organism with the nutrient had been deliberately "tailored" so as to make the lesser number of unknowns satisfy the greater number of equations.[6] Of course the interaction between living systems and nutrients is not random, but directed by the DNA molecule. Prebiotic systems, on the other hand, have no such endowment, and are subject to the problem of randomness alluded to by Wigner.

Landsberg[7] also used quantum mechanics to examine the question of spontaneous generation and reproduction of organisms. He found that by broadening Wigner's analysis to include nonequilibrium systems, the probabilities were greater than zero, though still very small. Based on the work of Wigner and Landsberg, we may conclude that quantum mechanics does not forbid the origin of life, but does suggest that life could not arise as a result of random interactions encountered in inanimate matter. The implication is that some hitherto little understood "principle of organization" must be responsible for the necessary "instructed" interaction of chemicals leading to the formation of living systems. This conclusion drawn from quantum mechanics is in agreement with the earlier observation from thermodynamics (Chapter 8) that a coupling of the energy flow through the system to the required work, especially configurational entropy work, is essential for the formation of life.

Polanyi[8] has emphasized that the mechanism and design in living organisms is irreducible to the laws of inanimate matter. He notes that the laws of chemistry and physics are expressed mathematically in terms of differential equations. The existence of living systems may only be understood, however, in the fixing of the boundary conditions that determine the form which both the equations and nature take. He leaves unanswered the question of how the "fixing" of the boundary conditions occurred, implying again the need for new laws.

In a similar way, Longuet-Higgins[9] affirms that physics and chemistry are conceptually inadequate as a theoretical framework for biology, and recommends more serious consideration of biological problems in terms of design, construction, and function.

The need for new laws is further underscored in the paradox seen by Schrodinger[10] in 1944. In inanimate matter, regular, orderly behavior is always the averaged result of the collective behavior of a large number of molecules acted on by particular constraints. In living systems, however, orderly behavior appears to result from the activity of single molecules or very small collections of molecules, in spite of the fact that fundamental physical laws lead us to believe that single molecules should behave in a random manner. Pattee[11] and Bohm[12] both have discussed this problem but have found no satisfactory solution. Bohm stresses that it is virtually certain that fundamental theory will not explain even the accurate transmission of genetic information, much less its origin. He further notes the ironic twist that just when physics and chemistry are abandoning mechanistic interpretations for probabilistic ones, biology is adopting them.

In summary, those who suggest new natural laws do not show it is reasonable to believe energy flow through a system would be coupled to accomplish the required work to produce the first protein, DNA, and ultimately, the first living cell. They simply point out that new organizing principles are *needed* as present ones are clearly inadequate. The mere need of new laws is a legitimate reason for seeking them, but only evidence can legitimately establish and sustain them. Intelligent contrivances harness a portion of the energy flow for work in the human world. How some energy converting/coupling means might arise without intelligence in the inorganic world before life is difficult to say.

Panspermia

Panspermia is the classical extraterrestrial view which originated after Pasteur's disproof of spontaneous generation in the 19th century, and was popularized earlier in this century by S. Arrhenius.[13] According to this view, a life spore was driven to earth from somewhere else in the cosmos by electromagnetic radiation pressure. The idea is sometimes called radiopanspermia.

Arrhenius calculated that if a particle were in the size range of 0.1-3 microns it could escape solar gravity and be pushed along in space by the pressure of light waves. Although Panspermia was an ingenious idea, it failed to account for three significant factors:

1. Panspermia did not really answer the question of origins; it merely pushed the problem to some other planet or place in the cosmos.
2. Panspermia offered no way to protect life spores from the lethal effects of intense radiation in space.

3. Panspermia offered no mechanism for safe entry through the earth's atmosphere. Arrhenius calculated that any life spore larger than 1 micron in diameter would burn up on entry. Most plant and animal cells, however, are in the range of 10-40 microns in diameter.

These problems were seen as severe and most people dismissed Panspermia as nonviable. Any flickering interest in Panspermia seemed to fade in the mid-Fifties with the emergence of the modern view of terrestrial abiogenesis.

Revived Interest in Panspermia

More recently, however, major objections to terrestrial chemical evolution scenarios, surveyed in the main body of the book, have caused some to reconsider Panspermia, even though it does not purport to be an account of life's origin. Why persist in looking to the earth for the answer to life's origin, especially since the evidence questioning terrestrial chemical evolution is quite substantial? As Brooks and Shaw noted, "We must be interested in the truths of matters and must not modify truths so that we can conveniently express our origins in ways which for some reason or other give us maximum satisfaction."[14]

Fred Hoyle and N.C. Wickramasinghe[15] have revived interest in Panspermia. They have offered calculations showing that particles up to 60 microns in size (which includes most living cells) could reach the earth, make "soft" landings, and neither burn up on entry in the atmosphere nor be obliterated on impact.

The problem of preserving life in space might not be as severe as Arrhenius thought. Through radioastronomy, organic molecules have been discovered in space, including some that are usually considered as precursors to life (e.g., formaldehyde, methanol), suggesting that some method of preservation is operative. Apparently these molecules are protected by thin layers of graphite dust a few tenths of a micron thick, which provide a shield from the destructive rays of ultraviolet light.

Added to this is the suggestive discovery of amino acids in meteorites, including some that are important in proteins. The Murchison meteorite, which fell in Australia in 1969, contained dl-amino acids,*

*More recent reports challenge this interpretation. See Michael H. Engel and Bartholomew Nagy, 1982. *Nature* **296**, 837.

including some proteinous ones. The presence of dl-amino acids was considered proof of extraterrestrial origin, and evidence that the meteorite was free of contamination from earth life. This is significant because the meteorite fell on a sheep farm, where remaining uncontaminated would be no trifling feat!

Perhaps more significant is the discovery of amino acids in another meteorite said to be 3.83 billion years old in the deep freeze of Antarctica. It was hailed as proof the amino acids were of extraterrestrial origin. According to Cyril Ponnamperuma, who conducted much of the investigation:

> The processes of chemical evolution appear to be common in the solar system ...Nobody has found life beyond the earth, but all of the evidence we are finding seems to point in that direction. I am certain that it is there.[16]

In spite of the optimism of Ponnamperuma and others, the relevance of these molecules from (and in) space is far from clear. Perhaps a story will put the issue in perspective. It is said that a little boy asked his mother whether it was true that we come from dust and at death we return to dust. After an affirmative reply from his mother the boy exclaimed, "Well, somebody is under my bed, but I can't tell if he's coming or going!" Only by a mistaken presupposition did the boy infer "somebody" from the presence of dust under his bed. The situation of the molecules in space seems remarkably parallel to this story. Clearly what is guiding some scientists to infer life in space from the mere presence of organic molecules is their hypothesis that life is rather common in the cosmos, being merely a stage in the development of matter. What else could have informed Ponnamperuma when he said concerning the possibility of life in space, "I am certain that it is there?" But surely the question is whether this hypothesis is correct; it is not an axiom for making a deduction.

We cannot disagree that there is need for an alternative to chemical evolution. In recognition of the fact that Panspermia offers no theory of origins, it must implicitly assume chemical evolution on some other locale in the cosmos, where conditions are more favorable than on earth. Many of the objections raised concerning terrestrial chemical evolution must, however, apply to other planets by the principle of uniformity. In any setting it comes down to the fact that natural forces acting alone must be capable of supplying the necessary configurational entropy work of building the protein, DNA, etc., and then assembling the cell. We know by experience that intelligent investigators can synthesize proteins and build genes.

We still have no evidence it can be done by unassisted abiotic means.

If one takes the view that only the organic materials from which to assemble life, and not life itself, came from space, then the next step must be faced. The assembly of life under these circumstances must occur in spite of the destructive forces discussed in Chapter 4. Space-incident organic molecules do little to solve the mystery of life's origin. As was pointed out in Chapter 4, two great concerns in order to have proper conditions for assembly of life are: (1) a source of precursor molecules and (2) protection of these till assembly occurs.

In spite of the problems with Panspermia, the number of scientists ready to defend it is growing.

Directed Panspermia

Also to be considered is an enterprising variation of Panspermia called Directed Panspermia.[17] Suggested by Francis Crick and Leslie Orgel, this hypothesis purports that life spores were sent to earth in some kind of rocket ship by extraterrestrial intelligence (ETI), most likely from some other galaxy. Speculations have been numerous. Perhaps ETI purposefully sent life spores to earth to make it a "wilderness area or zoo,"[18] or perhaps a cosmic dump site.[19] It is even possible life spores were left here inadvertently "on some ancient astronaut's boot."[20]

Like Panspermia, few scientists have considered Directed Panspermia worthy of pursuit. According to A. Dauvillier (who wrote prior to Directed Panspermia, but whose words are still appropriate),

> The doctrine of cosmic Panspermia can only be conceived if one accepts the idea of the carriage of live germs by foreign astronauts. This, to all intents and purposes, is a facile hypothesis, a subterfuge which seeks to avoid the fundamental problem of the origin of life.[21]

Most scientists probably agree with Dauvillier, that the notion of Panspermia directed by ETI is fantasy. There is some limited circumstantial evidence, however, that enhances its appeal over Panspermia. Like Panspermia, this view notes that there are some significant problems with terrestrial chemical evolution, such as the accumulating evidence for an oxidizing early atmosphere in contrast to the expected reducing condition. As Crick has mentioned, if it were really true that the primitive atmosphere contained a significant amount of oxygen, it would be difficult to imagine chemical evolution. In such a case, reasons Crick, "it would support the idea of Directed Panspermia."[22]

A fact that has enamored Crick is that in the fossil record, the earliest organisms appear suddenly without any evidence of a prebiotic soup or simple precursors.[23] For Crick this too is good evidence for Directed Panspermia. There is no compelling evidence that Crick and others can cite for this view, however. In fact the evidence cited above for Directed Panspermia would also apply to Panspermia. It is not surprising then to hear Crick lament, "Every time I write a paper on the origin of life, I swear I will never write another one, because there is too much speculation running after too few facts..."[24]

An additional form of "evidence" that is often used to support ideas about the existence of ETI in the cosmos is the Green Bank-Drake equation.[25] This equation gives the value of N, the number of advanced civilizations which are presently (presumably) communicating in the galaxy, to be

$$N = Rf_p n_e f_l f_i f_c f_d L$$

where R is the rate of star formation; f_p, the probability that a star will have planets; n_e, the number of planets per star with environments favorable to life; f_l, the probability that life will develop; f_i, the probability that intelligent life will develop; f_c, the probability that intelligent beings attempt interstellar communication; f_d, the probability that such beings desire to communicate; and L, the lifetime of a civilization after it reaches the interstellar communication stage.

Various estimates have been reached using the Green Bank-Drake equation. They range from N = 1 (even this value is assigned on the nearly universal assumption that spontaneous chemical evolution occurred once) to 10^8 or more. The wide spectrum of numbers cited in the literature for N reflects the room for individual subjectivity by those doing the estimating. Many enthusiasts consider it reasonable to conclude that perhaps a million advanced societies inhabit the cosmos. Several federally-funded projects such as Project OZMA, have been undertaken to search for extra-terrestrial intelligence (SETI). There is a growing body of literature critical of the ETI concept, however. For example, Frank Tipler has thoroughly examined the arguments for ETI, and notes, "the problem with the Drake equation is that only f_p, and to a lesser degree n_e, are subject to experimental determination."[26] Even when assigning to each term the value usually given in discussions of interstellar communities, the conclusion is reached that "we are alone."[27]

Directed Panspermia, like Panspermia itself, fails to give an account of the origin of life. It merely assumes that spontaneous generation must have occurred in some favored environment somewhere in the cosmos. Directed Panspermia is primarily a suggested mechanism to get life safely to earth. Surely intelligent beings could design an appropriate spaceship.

In spite of the lack of any real evidence, there appears to be growing interest in ETI among some scientists.

Special Creation by a Creator Within the Cosmos

Hoyle and Wickramasinghe[28] have developed a novel and creative argument, which we shall present in some detail. As will be seen, the view of intelligence creating biological specificity comes in not one, but two types: (1) a creating intelligence within the cosmos, and (2) a creating intelligence beyond the cosmos. In arguing for the former, Hoyle and Wickramasinghe contend that Darwinism has failed to account for the origin of life and the development of terrestrial biology.

> No matter how large the environment one considers, life cannot have had a random beginning...there are about two thousand enzymes, and the chance of obtaining them all in a random trial is only one part in $(10^{20})^{2000} = 10^{40,000}$, an outrageously small probability that could not be faced even if the whole universe consisted of organic soup.
>
> If one is not prejudiced either by social beliefs or by a scientific training into the conviction that life originated on the Earth, this simple calculation wipes the idea entirely out of court...the enormous information content of even the simplest living systems...cannot in our view be generated by what are often called "natural" processes, as for instance through meteorological and chemical processes occurring at the surface of a lifeless planet.... For life to have originated on the Earth it would be necessary that quite explicit instruction should have been provided for its assembly.... There is no way in which we can expect to avoid the need for information, no way in which we can simply get by with a bigger and better organic soup, as we ourselves hoped might be possible a year or two ago.[29]

The logic so far is that the customary notion of life originating by chemical evolution in an organic chemical soup is too improbable. The information content of living cells is too great to expect it to have arrived by "natural" means.

An adequate theory of origins requires an information source capable of generating chemical complexity. Hoyle and Wickramasinghe argue that the evidence is overwhelming that intelligence provided the information and produced life.

> The correct position we think is...an intelligence, which designed the biochemicals and gave rise to the origin of carbonaceous life.... Given an atlas showing the amino acid sequences of all the enzymes, human biochemists could construct them with complete accuracy, thereby demonstrating the enormous superiority of intelligence allied to knowledge over blind random processes.... Any theory with a probability of being correct that is larger than one part in $10^{40,000}$ must be judged superior to random shuffling. The theory that life was assembled by an intelligence has, we believe, a probability vastly higher than one part in $10^{40,000}$ of being the correct explanation of the many curious facts discussed in preceeding chapters.... Paley likened the precision of the living world to a beautifully made watch. He then argued that, just as a watch owes its origin to a watchmaker, the world of Nature must owe its origin to a Creator, *God*.... The speculations of *The Origin of Species* turned out to be wrong.... It is ironic that the scientific facts throw Darwin out, but leave William Paley, a figure of fun to the scientific world for more than a century, still in the tournament with a chance of being the ultimate winner.... Indeed, such a theory is so obvious that one wonders why it is not widely accepted as being self-evident. The reasons are psychological rather than scientific.[30]

To be sure, such a creative view entails purpose, a point which Hoyle and Wickramasinghe address.

> The revulsion which biologists feel to the thought that purpose might have a place in the structure of biology is therefore revulsion to the concept that biology might have a connection to an intelligence higher than our own.[31]

By this time surely every schoolboy has figured out that Hoyle and Wickramasinghe are offering to the world the traditional view of Special Creation. But every schoolboy would be wrong! Hoyle and Wickramasinghe deny the creator is the traditional supernatural God. They envision a creator within the total cosmos. They contend that a flaw in logic kept generations of scientists from seeing the truth that intelligence is the authentic source of the information in the biological world.

> The whole of the special creation theory was thought to be wrong and there was a general revulsion among scientists against it. In effect, because the details were seen to be incorrect, the fundamental idea that life was created by an intelligence was also rejected.... If we define "creation" to mean arrival at the Earth from outside, the unit of creation in our picture is the gene, not the working assembly of genes that we call a species.[32]

The novelty of this suggestion is that is seems to solve the major problem of the origin of life that both Panspermias merely skirted. A real origin is suggested, primarily of genes but also of some bacterial cells. The implication is that the mechanism of Panspermia can be used to safely transport these genes to earth without having to resort to anything as elaborate as a spaceship. Since genes or gene fragments would be within the size range of 0.1-3 microns, light waves could easily move them across the solar system. Furthermore, they could be protected from intense radiation in space by a thin sheath of graphite. Finally, they would be well within the 60 microns limit for safe entry into our atmosphere without burning up.

Not only would such a process as this operate at the beginning, Hoyle and Wickramasinghe suggest it is a continuous process through history even to this day.

> In our view the arrival at the Earth of living cells, and of fragments of [created] genetic material more generally, is a continuing ongoing process that directs the main feature of biological evolution. It is this process which does the job that is usually attributed to Darwinism.[33]

In addition to the origin of life, Hoyle and Wickramasinghe account for the whole of biology by these falling genes. The gaps in the fossil record are real; there never were transitional forms, because the genetic information necessary for the jumps in species came continuously to earth by cosmic means.

If the cosmic intelligence responsible for the creation of genes and bacteria is not God, then who or what is it?

> The advantage of looking to the whole universe is rather that it offers a staggering range of possibilities which are not available here on the Earth. For one thing it offers the possibility of high intelligence within the universe that is not God. It offers many levels of intelligence rising upwards from ourselves.... To be consistent logically, we have to say that the intelligence which assembled the enzymes did not itself contain them. This is tantamount to arguing that carbonaceous life was invented by a noncarbonaceous intelligence, which by no means need be *God*, however.[34]

What other kind of high intelligence is also free of enzymes? The answer offered is a *philosophical entity*. In order to solve the problem of the origin of life, Hoyle and Wickramasinghe seem to have relied on Hindu philosophy for their views. A similar view was expressed by Plotinus,* the father of neo-Platonism. In both views

---

*See Plotinus, *Enneads*, trans. by Stephen MacKenna, 1966. London: Faber and Faber.

the creator is part of the universe, being subordinate to the ultimate reality (Brahman, or the One which is beyond the universe, and unknowable).

Hoyle and Wickramasinghe speculate further that the intelligence may not have simply remained in the outer regions of the cosmos, but may have in fact become incarnate on earth in a sort of "invasion from space."

> We come now to what for us is a strong argument for the existence of an overt plan of planetary invasion...we have so far been unable to exterminate a single insect species.
> Not even one species among millions![35]

And what do we learn from this curious fact?

> The situation points clearly to one of two possibilities. Either we are dealing with an overt plan invented by an intelligence considerably higher than our own,...or the insects have already experienced selection pressure against intelligences of at least our level in many other environments elsewhere in the universe.[36]

The moment of truth finally arrives when we learn the identity of the superintelligence. Hoyle and Wickramasinghe ask, "Could the insects themselves be the intelligence higher than our own?"[37] If anyone wonders why we are so long discovering their true identity, Hoyle and Wickramasinghe suggest it is because they do not wish to be known.

> Perhaps concealment is an essential tactic. Perhaps the intelligence is static because it understands the dictum of sagacious lawyers: "When your case is going well, say nothing."[38]

We suspect that few will find Hoyle and Wickramasinghe's hypothesis of falling genes acceptable as a genuine contribution of science. Although their criticism of chemical evolution is cogent, the novel notion of cosmically created genes falling to the earth does not realistically take into account the fate of genes once they reach the earth (Chapter 4) nor heed the fact that genes need a proper cellular context in which to work, nor allow that the configurational entropy work requirement applies to cell assembly too.

Genes are complex segments of DNA. As we saw in Chapter 4, they are extremely vulnerable to a host of chemicals that surely would have been present under reducing conditions. On the other hand, oxidizing conditions would have been even worse for gene survival. Genes are wonderful templates for building enzymes, but without a cellular host endowed with the appropriate enzymes they are powerless to do synthesis. One could perhaps so contrive the surrounding milieu in a laboratory setting that cellular conditions are mimicked to bring about replication and enzyme building. Such a possibility is extremely doubtful in the prebiotic world—even one rained upon by cosmic genes from above.

Special Creation by a Creator Beyond the Cosmos

In agreement with views of abiogenesis, and the foregoing view of Hoyle and Wickramasinghe, Special Creation by a Creator beyond the cosmos holds there was once a time in the past when matter was in a simple arrangement, inert and lifeless. Then at a later time matter was in the state of biological specificity sufficient for bearing and sustaining life. Special Creation (whether from within the cosmos or beyond it) differs from abiogenesis in holding that the *source which produced* life was intelligent.

Throughout history, many writers have attempted to describe the work of the Creator. What they all seem to hold in common is the idea that an intelligent Creator *informed* inert* matter by shaping it as a potter fashions clay. Some representations are quite anthropomorphic, others less so. But there is considerable agreement that somehow an active intellect produced life.

In 1967 J.D. Bernal, a leading developer of the chemical evolution scenario, issued a challenge to divine creationists. He said:

> Now that we are embarking on a serious scientific discussion on the origin of life, it is time...we were furnished with a more precise, complete and self-consistent account of the spiritual or divine origin of life than any that have been produced as an alternative to the mechanistic one. Such an argument, ...should provide us with a *clearer path to further scientific advance*, even if it does not reach the end.[39] (Emphasis added.)

We do not believe there has been any significant response to Bernal's challenge that would "provide us with a clearer path to further scientific advance." In fact, what follows should be viewed as only introductory to that end.

---

*It must be acknowledged that the idea of inert matter did not arrive in its modern understanding until the scientific revolution of the 16th and 17th centuries. This matters little, however, with reference to first life. For as noted in Chapter 2, the Church considered spontaneous generation only as a secondary origin.

What Concerns Scientists About Creation?

(1) Creation involves the supernatural. It is common knowledge that the claim that an active intellect informed nature has been on uneasy terms with the mainstream of science. To anyone trained in science, the reason is no mystery. It involves the supernatural. The objection is expressed well by the recognized science writer, J.W.N. Sullivan. Upon his death Sullivan was described by Time magazine as "one of the world's four or five most brilliant interpreters of physics to the world of common men."[40] He showed the concern most scientists have in considering a theistic explanation of the origin of life. Sullivan said (in 1933, but still cogent today):

> The beginning of the evolutionary process raises a question which is as yet unanswerable. What was the origin of life on this planet? Until fairly recent times there was a pretty general belief in the occurrence of "spontaneous generation." ...But careful experiments, notably those of Pasteur, showed that this conclusion was due to imperfect observation, and it became an accepted doctrine that life never arises except from life. So far as actual evidence goes, this is still the only possible conclusion. *But since it is a conclusion that seems to lead back to some supernatural creative act, it is a conclusion that scientific men find very difficult of acceptance.*[41] (Emphasis added.)

So it is the supernatural that concerns many scientists. But what is it about the supernatural that troubles them? Why is creation difficult to accept?

(2) Creation entails discontinuity. A major concern of many scientists is that to allow supernatural involvement is to introduce *discontinuity* into science. Continuing to quote Sullivan:

> It carries with it what are felt to be, in the present mental climate, undesirable philosophic implications, and it is opposed to the scientific desire for continuity. *It introduces an unaccountable break in the chain of causation*, and therefore cannot be admitted as part of science unless it is quite impossible to reject it. For that reason most scientific men prefer to believe that life arose, in accordance with the laws of physics and chemistry.[42] (Emphasis added.)

Here is the vision of nature as a seamless web of causal connections, an idea dominant in science for more than 250 years. As Einstein wrote, "The scientist is possessed by the sense of universal causation."[43] And, of course, creation would be a discontinuity. Hans Gaffron also expressed this concern in his address to the Darwin Centenniel Celebration in 1959. Regarding chemical evolution Gaffron said:

> [it] is a nice theory, but no shred of evidence, no single fact whatever, forces us to believe it. What exists is only the scientist's wish *not to admit a discontinuity in nature* and not to assume a creative act forever beyond comprehension.[44] (Emphasis added.)

Notice, however, that in the above quotations of Sullivan, Einstein, and Gaffron there is only a *desire*, *sense*, *preference*, and *wish* that nature be continuous. This is important to understand because the wish went unfulfilled. The great quantum revolution has banished the notion of continuity as a necessity in science. According to de Broglie, one of the pioneers of the new physics, "on the day when quanta, surreptitiously, were introduced the vast and grandiose edifice of classical physics found itself shaken to its very foundation."[45] In addition advances in astronomy, as chronicled by Robert Jastrow,[46] have made it clear there was also a discontinuity at the beginning of the world. In fact there seems to be no good reason to suppose an original discontinuity would undermine a scientific understanding of the functioning of the world. For science in this sense is not concerned with the origin but with the operation of the world. It is clear from these developments in science that discontinuity is not the whole reason that creation is difficult for many scientists to accept.

(3) Creation might destroy the scientific quest for knowledge. Even though the structure of science and scientists themselves have survived the news that at bottom reality is discontinuous, there is no less suspicion that creation would stifle the quest for knowledge. But would creation necessarily destroy the scientific quest and hence bring an end to science?

In giving answer to this question it will be necessary to briefly consider the nature of science.

Operation Science and the God Hypothesis

It is widely appreciated that from its beginning modern science has been concerned with finding and describing orderly pattern in the recurring events of nature. To do this a well-defined method is used. Data are gathered through observation and experimentation. As data are gathered, theories are proposed to explain the behavior or operation of the phenomena investigated. According to wide usage, a valid theory of science must pass a three-fold test:[47]

1) Its ability to explain what has been observed.

2) Its ability to explain what has not yet been observed.
3) Its ability to be tested by further experimentation and to be modified as required by the gathering of new data.

Notice, however, that this approach to testing theories only works if there is some pattern of recurring events against which theories can be checked and falsified if they are false. Through repeated observation attention is focused on a class of events, each of which is similar. The equations describing the behavior of the class would be applicable to any of its individual members. Let us say, for example, we have a theory about earth orbiting the sun and we propose to test it by predicting a solar eclipse. Although a particular eclipse would be the focus of the experiment, the result would apply to solar eclipses as a general class. Because there are recurring patterns of celestial movements we can test the theory. Such theories are operation theories. That is, they refer to the ongoing operation of the universe. We shall call the domain of operation theories *operation science* for these theories are concerned with the recurring phenomena of nature. Examples of operation science include the recurring motion of planets about the sun, the swinging of a pendulum, the parabolic trajectory of a cannonball, a single cell turning by stages into a fully formed organism, the recurrent cubic structure of table salt crystallizing out of water solution and the migration of a Monarch butterfly. These and many other phenomena have been accounted for in the language of operation science. Because of its familiarity and long, successful history, it is surely what most people think of when they think about science.

Here in operation science the appeal to God is quite illegitimate, since by definition God's supernatural action would be willed at His pleasure and not in a recurring manner. Yet it is true that on numerous occasions throughout the history of science there have been those who have appealed to the God-hypothesis to "solve" some knotty problem of the ongoing operation of the universe instead of grappling with it and searching for natural causes to explain it.

Basically the idea of the God hypothesis is that whenever there is a gap in our knowledge, we run God in as a "bit-player," so to speak, to fill the gap. This view is known fittingly as the God-of-the-gaps. There is legitimate concern about this means of solving problems in operation science.

A classic example of this approach to scientific problem-solving is seen in the life of the great Isaac Newton, who appealed to the God-hypothesis to account for certain anomalies in the heavens.

(Note that an anomaly was defined by reference to Newton's own view of things.) Later, Laplace accounted for such discrepancies in a perfectly lawful manner. This was an important but painful lesson for scientists to learn. The illustration is sharpened by the story of the French Emperor Napoleon who asked Laplace where God fit in his equations, to which Laplace responded, "Sire, I have no need of that hypothesis."[48] Although some have misunderstood Laplace's reply in this instance as being anti-God, it was quite appropriate.

Origin Science

On the other hand an understanding of the universe includes some singular events, such as origins. Unlike the recurrent operation of the universe, origins cannot be repeated for experimental test. The beginning of life, for example, just won't repeat itself so we can test our theories. In the customary language of science, theories of origins (*origin science*) cannot be falsified by empirical test if they are false, as can theories of operation science.

How then are origins investigated? The method of approach is appropriately modified to deal with unrepeatable singular events. The investigation of origins may be compared to sleuthing an unwitnessed murder, as discussed in Chapter 11. Such scenarios of reconstruction may be deemed plausible or implausible. Hypotheses of origin science, however, are not empirically testable or falsifiable since the datum needed for experimental test (namely, the origin) is unavailable. In contrast to operation science where the focus is on a class of many events, origin science is concerned with a particular event, i.e., a class of one.

When Galileo's ideas on acceleration (operation science) were presented, observers were not limited to mere plausibility. They could actually empirically falsify the claims of Galileo had they been false. Indeed Pasteur's falsification of spontaneous generation was possible only because it was said to recur in the domain of operation science. Appropriate testing against nature falsified the notion of spontaneous generation. The best we can ever hope to achieve with wrong ideas about origins is to render them *implausible*. By the nature of the case, true falsification is out of the question.

In spite of this fundamental difference between origin science and operation science, there is today very little recognition of it, and an almost universal convention of excluding the divine from origin science as well as from operation science. This has occurred without any careful prior analysis of the problem to see if the exclusion is valid in the case of origin science. It seems to have been merely assumed.

An example of this exclusion by assumption instead of valid argument comes from this statement by Orgel:

> Any "living" system must come into existence either as a consequence of a long evolutionary process or a miracle.... Since, as scientists, we must not postulate miracles we must suppose that the appearance of "life" is necessarily preceded by a period of evolution.[49]

We agree with Orgel that miracles must not be posited for operation science.* We disagree with Orgel however, and others, when it is merely assumed that the exclusion of the divine from origin science is valid. This has not been demonstrated.

There are significant and far-ranging consequences in the failure to perceive the legitimate distinction between origin science and operation science. Without the distinction we inevitably lump origin and operation questions together as if answers to both are sought in the same manner and can be equally known. Then, following the accepted practice of omitting appeals to divine action in recurrent nature, we extend it to origin questions too. The blurring of these two categories partially explains the widely held view that a divine origin of life must not be admitted into the *scientific* discussion, lest it undermine the motive to inquire and thus imperil the scientific enterprise. This is what Preston Cloud meant when he noted, "The most serious threat of creationism is that, if successful, it would stifle inquiry."[50] One can also see the same concern echoed by Stansfield:

> ...the creationist can easily explain any phenomenon by simply saying "God did it." This approach, though it may be perfectly correct in an absolute sense, does not foster further inquiry and is therefore intellectually emasculated.[51]

The perception of a threat to scientific inquiry and the possible end of science are legitimate concerns. But we question whether the God-hypothesis in origin science would necessarily have this disastrous effect. Just a little reflection on the history of science brings out the irony in the current state of affairs. For there is a rather impressive reason to doubt that science (i.e., operation science) would suffer much by positing Special Creation by a Creator beyond the cosmos.

*For a critical evaluation of the long-standing tendency to reject miracles in modern thought, see Norman L. Geisler, 1982, *Miracles and Modern Thought*. Grand Rapids, Michigan: Zondervan Publishing House, Chps. 1-8.

On the contrary, it turns out that this very idea of creation played a significant role in the origin of modern science. Speaking with one voice on this point are such diverse authors as Alfred N. Whitehead,[52] Melvin Calvin,[53] Michael B. Foster,[54] R. Hooykaas,[55] Loren Eisley,[56] C.F. von Weizsacker,[57] Stanley Jaki,[58] J. Robert Oppenheimer,[59] and Langdon Gilkey.[60] For example, Eisley said the birth of modern science was due to:

> The sheer act of faith that the universe possessed order and could be interpreted by rational minds.... The philosophy of experimental science...began its discoveries and made use of its method in the faith, not the knowledge, that it was dealing with a rational universe controlled by *a Creator who did not act upon whim nor interfere with the forces He had set in operation*. The experimental method succeeded beyond man's wildest dreams but the faith that brought it into being owes something to the Christian conception of the nature of God. It is surely one of the curious paradoxes of history that science, which professionally has little to do with faith, owes its origins to an act of faith that the universe can be rationally interpreted, and that science today is sustained by that assumption.[61] (Emphasis added.)

Notice that while Eisley does not identify the distinction between operation science and origin science, the distinction is implicit in his explanation that a great deal of good science was done by early modern scientists who allowed at least a few discontinuities, i.e., the origin of matter, universe, life.

It would be quite ironic if the very idea of creation which provided much of the energy and impetus to launch modern natural science (and did so without noticeable lethargy) should lead to the demise of this same science. In our view, as long as one acknowledges and abides by the above distinction between origin science and operation science, there is no *necessary* reason that Special Creation would have the disastrous effects predicted for it. One must be careful, however, to follow the tradition of early modern scientists and *disallow* any divine intervention in operation science.

Why then is Special Creation so summarily dismissed by nearly all writers, especially since it is typically listed as a theoretical alternative for the origin of life? Our analysis suggests that failure to properly distinguish origin science and operation science has led many to dismiss creation. Also we believe another factor is involved, and is worthy of discussing in some detail. To be sure, the matter of discontinuity, and the possible demise of science discussed above are part of the reason. But we should not ignore our own humanness, and the role of metaphysical thinking in the origin of life question.

Metaphysics and The Origin of Life

Hilde Hein, in her book *On the Nature and Origin of Life*, says that "a metaphysical position...makes a claim about reality which is somehow prior to or more fundamental than our scientific or common-sense observations."[62] How we happen to come by these metaphysical positions is of no concern to us here. However, as Hein continues,

> once it is adopted, it will shape, rather than be shaped by, our scientific and common-sense observations. This is to say that, on the whole our metaphysical commitment has priority over our scientific and common-sense beliefs such that, if challenged, they will yield to it rather than the reverse.[63]

It might appear that if metaphysical views have such control over us, the best approach would be simply to look at reality straight-on without any metaphysical lens at all. This, however, is not an option that is open to us. The grand old days of positivism, when people actually thought this possible, are over.

Scientific developments earlier in this century, particularly in the area of relativity and quantum physics, have shown presuppositionless science to be a myth. The powerful writings of Polanyi,[64] Popper,[65] Kuhn,[66] Toulmin,[67] and others have strictly shown that because of the role of the observer (e.g., actually disturbing the object during the act of observing) it is difficult for objective reality to be objectively known.

Old myths die hard, however. Although news of these advances in science and philosophy earlier in the century are filtering through society, their effect in some quarters is minimal and there are dangerous consequences as a result. As David Bohm has written:

> It seems clear that everybody has got some kind of metaphysics, even if he thinks he hasn't got any. Indeed, the practical "hard-headed" individual who "only goes by what he sees" generally has a very dangerous kind of metaphysics, i.e., the kind of which he is unaware.... Such metaphysics is dangerous because, in it, assumptions and inferences are being mistaken for directly observed facts, with the result that they are effectively riveted in an almost unchangeable way into the structure of thought.[68]

Bohm then adds some practical advice:

> One of the best ways of a person becoming aware of his own tacit metaphysical assumptions is to be confronted by several other kinds. His first reaction is often of violent disturbance, as views that are very dear are questioned or

> thrown to the ground. Nevertheless, if he will "stay with it," rather than escape into anger and unjustified rejection of contrary ideas, he will discover that this disturbance is very beneficial. For now he becomes aware of the assumptive character of a great many previously unquestioned features of his own thinking.[69]

We believe Bohm is quite right. It is in the interest of science to have the metaphysical assumptions out on the table. Just what are the fundamental metaphysical alternatives in the question of the origin of life? Historically, they have been called theism and naturalism. For simplicity, we will note that theism affirms a fundamental distinction between the Creator and the creature, while naturalism denies this absolute distinction and defines all of reality in terms of what theists see as some aspect of the created world.*

The origin perspective of metaphysical naturalism is spontaneous generation (abiogenesis), and of theism† it is Special Creation. It follows from what Bohm has said that a great deal of practical self-awareness of our individual views would probably emerge if we allowed ourselves to be confronted with both theism and naturalism in the area of origins. Very often the debate between theism and naturalism is cast as a conflict between religion (i.e., the supernatural) and science. However, as Ian Barbour has pointed out this is a mistake. It is "a conflict between two metaphysical interpretations of the nature of reality and the significance of human life."[70]

Metaphysical Commitment vs. Unreason

If metaphysical positions have such a controlling influence as Hein has indicated, this raises a practical question. In the face of contradictory evidence, when is one to be praised for metaphysical commitment, and chided for unreasonable faith? The answer one gives to this question depends in large measure on the metaphysical stance already adopted. To illustrate, consider George Wald's discussion of how biologists responded after Pasteur's refutation of spontaneous generation. Says Wald:

---

*Western naturalism has typically defined the world in material terms while Eastern naturalism has emphasized the spiritual. What marks out both of these great traditions as naturalistic is that both deny an *absolute* Creator who is really distinct from creation, even though, as we saw with the view of Hoyle and Wickramasinghe, a creator within the universe has sometimes been posited.

†Distinctions within theism are beyond the scope of the present work.

> We tell this story [of Pasteur's experiments] to beginning students of biology as though it represents a triumph of reason over mysticism. In fact it is very nearly the opposite. The reasonable view was to believe in spontaneous generation; the only alternative, to believe in a single, primary act of supernatural creation. There is no third position.[71]

Wald is saying that there are times when it is clearly *unreasonable* to follow the evidence where it leads. When? Those times when following the evidence would lead one to the supernatural. This is an example of metaphysical commitment to naturalism in the face of contradictory evidence. Clair E. Folsome[72] represents another example of commitment to metaphysical naturalism in spite of contradictory evidence. Folsome critiqued the abiogenesis that Wald had upheld. Folsome pointed out the extreme dilution of the primitive soup, the scarcity of organic nitrogen in the early sediments, and the grave deficiencies in the concentration mechanism proposed for the primitive water basins. He then noted: "Every time we examine the specifics of the theories presented by Oparin and Bernal, current information seems to contradict them."[73] Does Folsome then entertain doubt as to the plausibility of the Oparin-Bernal hypothesis? No.

This also is apparently a time when it would be *unreasonable* to follow the evidence where it leads. Instead, Folsome expresses his commitment, "yet, in the main, *they were right* [in postulating that some sort of chemical evolution had occurred]...their models were wrong, but the central theme they pursued seems even more right now than before."[74] (Emphasis added.)

Of course, creationists also manifest a similar commitment to theism, even if like Wald and Folsome they remain silent about their metaphysical stance. We have not bothered to document this for theism, since it is generally acknowledged.

Special Creation and the Evidence

Special Creation by a Creator beyond the cosmos envisions a prepared earth with oxidizing conditions, an earth ready to receive life. It is suggestive then that there has been accumulating evidence for an oxidizing early earth and atmosphere. If the early earth were really oxidizing it would not only support creation, it would also be difficult to even imagine chemical evolution. Similarly, the short time interval (< 170 my) between earth's cooling and the earliest evidence of life supports the notion of creation. And, of course, if life were really created it would account for there being so little nitrogen in Precambrian sediments (there never was a prebiotic soup). In addition, Special Creation accords well with the observed boundary between what has been done in the laboratory by abiotic means and what has been done only through interference by the experimenter. If an intelligent Creator produced the first life, then it may well be true that this observed boundary in the laboratory is real, and will persist independent of experimental progress or new discoveries about natural processes. Also an intelligent Creator could conceivably accomplish the quite considerable configurational entropy work necessary to build informational macromolecules and construct true cells. As Fong has said:

> The question of the ultimate source of information is not trivial. In fact it is the basic and central philosophical and theoretical problem. The essence of the theory of Divine Creation is that the ultimate source of information has a separate, independent existence beyond and before the material system, this being the main point of the Johannine Prologue.[75]

It is doubtful that any would deny that an intelligent Creator *could* conceivably prepare earth with oxidizing conditions and create life. And, of course, the data discussed above are consistent (and compatible) with this view of Special Creation. What we would like to know, of course, is whether an intelligent Creator *did* create life. The question, unfortunately, is beyond the power of science to answer. Another question which can be answered, however, is whether such a view as Special Creation is plausible.

Plausibility and Creation

On several occasions we have indicated that hypotheses of origin science may be evaluated in terms of their plausibility, but falsification, the language of operation science, will not apply. How then does one determine whether an origin science scenario is plausible? The principles of causality and uniformity are used. *Cause* means that necessary and sufficient condition that alone can explain the occurrence of a given event. By the principle of *uniformity* is meant that the kinds of causes we observe producing certain effects today can be counted on to have produced similar effects in the past. We can go back into the past with some measure of plausibility only by assuming the kind of cause needed to produce that kind of effect in the present was also needed to produce it in the past. In other words, "the present is a key to the past."

As we saw, this is how scientists have arrived at the reconstructed scenario of a prebiotic earth. What makes views of abiogenesis legitimate as origin science then is the assumed legitimacy of cause-effect reasoning and the principle of uniformity.

The dilemma for chemical evolution, however, has been failure to identify any contemporary example of specified complexity (as distinct from order, see Chapter 8) arising by abiotic causes. What is needed is to identify in the *present* an abiotic cause of specified complexity. This would then provide a basis for extrapolating its use into the past as a conceivable abiotic cause for supplying the configuration entropy work in the synthesis of primitive DNA, protein, and cells. The failure to identify such a contemporary abiotic cause of specified complexity is yet another way to support our conclusion that chemical evolution is an implausible hypothesis.

But does creation employ cause-effect and the principle of uniformity? Yes. In fact, it appeals to them as the only way we can plausibly reconstruct the past. Consider, for example, the matter of accounting for the informational molecule, DNA. We have observational evidence in the present that intelligent investigators can (and do) build contrivances to channel energy down nonrandom chemical pathways to bring about some complex chemical synthesis, even gene building. May not the principle of uniformity then be used in a broader frame of consideration to suggest that DNA had an intelligent cause at the beginning? Usually the answer given is no. But theoretically, at least, it would seem the answer should be yes in order to avoid the charge that the deck is stacked in favor of naturalism.

We know that in numerous cases certain effects always have intelligent causes, such as dictionaries, sculptures, machines and paintings. We reason by analogy that similar effects also have intelligent causes. For example, after looking up to see "BUY FORD" spelled out in smoke across the sky we infer the presence of a skywriter even if we heard or saw no airplane. We would similarly conclude the presence of intelligent activity were we to come upon an elephant-shaped topiary in a cedar forest.

In like manner an intelligible communication via radio signal from some distant galaxy would be widely hailed as evidence of an intelligent source. Why then doesn't the message sequence on the DNA molecule also constitute *prima facie* evidence for an intelligent source? After all, DNA information is not just analogous to a message sequence such as Morse code, it *is* such a message sequence.[76] The so-called Shannon information laws apply equally to the genetic code and to the Morse code. True, our knowledge of intelligence has been restricted to biology-based advanced organisms, but it is currently argued by some that intelligence exists in complex non-biological computer circuitry. If our minds are capable of imagining intelligence freed from biology in this sense, then why not in the sense of an intelligent being before biological life existed?[77]

We believe that if this question is considered, it will be seen that most often it is answered in the negative simply because it is thought to be inappropriate to bring a Creator into science.

The above discussion is not meant as a scientific proof of a Creator, but is merely a line of reasoning to show that Special Creation by a Creator beyond the cosmos is a plausible view of origin science.

Metaphysical Tolerance: A Discipline for Progress

To be sure, there are sensitive issues involved when we begin to explore the metaphysical questions surrounding the origin of life. However, there is no easy way to resolve these issues. The only sure path is difficult. It demands the discipline required to temporarily table our personal tastes and preferences and humble ourselves in order to give serious consideration to how the data can be viewed from the other metaphysical position. We must do so recognizing that the truth of origins surely remains the truth regardless of which metaphysical position we individually adopt. As Melvin Calvin has observed, "The true student will seek evidence to establish fact rather than confirm his own concept of truth, for truth exists whether it is discovered or not."[78] The difficulty in pursuing these metaphysical matters is that scientists on the whole have seen so little value in this pursuit. After the birth of modern science in the 17th century it became an accepted procedure by the end of the 19th century to separate science and metaphysics into isolated, thought-tight compartments. This seemed to work well in practice, for after science got started the practitioners of science could function without even being aware of the metaphysical basis on which they operated. The modern scientific tradition has largely developed within the area we have called operation science, with its emphasis on recurring phenomena and testable hypotheses. Because of the inertia of heritage, the practice of science continued with only a few practicing scientists apparently aware of its metaphysical basis. As a result, now that we need to negotiate metaphysical terrain for proper understanding of origin science, few in science are equipped with the requisite skills. We believe this is a major reason creation in the area of origin science is viewed with such deep suspicion by many and simply dismissed.

When we are asked to consider "far out" or "strange" ideas such as Special Creation, as were the authors just a few years ago, typically the response is exactly that mentioned by Bohm as cited earlier. "His first reaction is often of violent disturbance." This was our reaction, too. However, as Bohm goes on to say, if one is willing to "stick with the inquiry rather than escape into anger or unjustified rejection of contrary ideas...he becomes aware of the assumptive character of a great many previously unquestioned features of his own thinking."

The process as Bohm described it can sometimes be painful (it was to one of the authors) but the quest for truth has never been easy, and has on more than a few occasions been known to make one unpopular.

To be sure, not everyone who goes into the matter will reach the creationist conclusion that we have. Even so, in the words of Davis and Solomon, as expressed in their book *World of Biology*:

> We cannot imagine that the cause of truth is served by keeping unpopular or minority ideas under wraps... Specious arguments can be exposed only by examining them. Nothing is so unscientific as the inquisition mentality that served, as it thought, the truth, by seeking to suppress or conceal dissent rather than by grappling with it.[79]

As with the court trial by jury analogy discussed in Chapter 11, we believe both sides[80] of the origins issue (i.e., representatives of both metaphysical categories) must be considered, precisely because there is no way to test origins ideas in origin science against recurring phenomena (origins by definition do not recur). The issue will be decided on the basis of plausibility, not falsifiability. There is good historical precedent for this approach. Charles Darwin in his introduction to *The Origin of Species* said:

> For I am well aware that scarcely a single point is discussed in this volume on which facts cannot be adduced, often apparently leading to conclusions directly opposite to those at which I have arrived. *A fair result can be obtained only by fully stating and balancing the facts and arguments on both sides of each question*, and this is here impossible.[81] (Emphasis added.)

Presenting origin science ideas from both metaphysical categories—

theism and naturalism—in addition to giving an opportunity to choose the most plausible view from the total theoretical spectrum, will also help us become aware of:

- our own position and why we hold it
- the weaknesses and disadvantages of our position
- the need for tolerance of others' positions and
- the limitations of science.

Our purpose in this epilogue has been to shed light on the issues and to avoid heat as much as possible. Only the reader can judge how successful we have been. If there is but one thing our acquaintance with the history of science has taught us, it is that unless some progress is made in recognizing the role of metaphysical thinking and properly using it, the origins debate will simply rage on, much as it has in the past, with representatives of each side of the dispute failing to hear or understand the other. Consequently, such scientists who go along blithely oblivious to the role of metaphysical thinking will simply act as if data really are observed and comprehended as neutral fact. Hopefully the lion of positivism has made its last roar and we can learn from advances in philosophy and science since the time of Darwin. If we can learn from our mistakes, we may expect more productive interchanges in the future. Toward that end we reach.

References

1. W.M. Elsasser, 1958. *Physical Foundation of Biology*. New York and London: Pergamon Press; 1966. *Atom and Organism*. Princeton: Princeton University Press.
2. M.A. Garstens, 1969. In C.H. Waddington ed., *Towards a Theoretical Biology*, "Statistical Mechanics and Theoretical Biology," **2**, 285, and "Remarks on statistical mechanics and theoretical biology," **3**, 167. Edinburgh: Edinburgh University Press.
3. P.T. Mora, *Nature* **199**, 212.
4. E.P. Wigner, 1961. "The Probability of the Existence of a Self-Reproducing Unit," in *The Logic of Personal Knowledge*, essays presented to M. Polanyi, ed. Edwards Shils. London: Routledge and Kegan Paul, p. 231.
5. J. von Neumann, 1932. *Mathematische Grundlagen der Quantenmechanik*. Berlin: Julius Springer, Berlin, (English translation: 1955. Princeton: Princeton University Press, Chapter 5.)
6. Wigner, p. 235.

7. P.T. Landsberg, 1964. *Nature* **203**, 928.
8. M. Polanyi, Aug. 21, 1967. "Life Transcending Chemistry and Physics," *Chem. Eng. News*, p. 54.
9. C. Longuet-Higgins, 1969. "What Biology is About," in C.H. Waddington ed., *Towards a Theoretical Biology*, **2**, 227, Edinburgh University Press.
10. E. Schrodinger, 1945. *What is Life*? London: Cambridge University Press, and New York: MacMillan.
11. H.H. Pattee, in C.H. Waddington ed., *Towards a Theoretical Biology* "The Physical Basis of Coding and Reliability in Biological Evolution," **2**, 268. "The Problems of Biological Hierarchy," **3**, 117. "Laws and Constraints, Symbols and Languages," **4**, 248.
12. D.Bohm, 1969. "Some Remarks on the Notion of Order," in C.H. Waddington ed., *Towards a Theoretical Biology* **2**, 18, Edinburgh: University Press.
13. S. Arrhenius, 1908. *Worlds in the Making*. New York: Harper and Row.
14. J. Brooks and G. Shaw, 1973. *Origin and Development of Living Systems*. London and New York: Academic Press, p. 360.
15. F. Hoyle and N.C. Wickramasinghe, 1978. *Lifecloud*. London: J.M. Dent; 1979. *Diseases From Space*, London: J.M. Dent; 1981. *Space Travelers: The Bringers of Life*, Cardiff: University of Cardiff Press; 1981. *Evolution From Space*, London, J.M. Dent.
16. C. Ponnamperuma, January 10, 1979. Quoted in "Odds Favor Life Beyond the Earth," Knight-Ridder News Wire; *Dallas Times Herald*, B3.
17. F.H.C. Crick and L.E. Orgel, 1973. *Icarus* **19**, 341.
18. J.A. Ball, 1973. *Icarus* **19**, 347.
19. T. Gold, May 1960. *Air Force and Space Digest*, p. 65.
20. Brooks and Shaw, p. 360.
21. A. Dauvillier, 1965. *The Photochemical Origin of Life*. New York: Academic Press, p. 2.
22. Francis Crick, 1981. *Life Itself*. New York: Simon and Schuster, p. 79.
23. Ibid., p. 144.
24. Ibid., p. 153.
25. Robert T. Rood and James S. Trefil, 1981. *Are We Alone?*. New York: Charles Scribners and Sons; I.S. Shklovskii and Carl Sagan, 1966. *Intelligent Life in the Universe*. San Francisco: Holden Day; Frank Tipler, April 1981. *Physics Today* **34**, No. 4, p. 9.
26. Tipler, p. 71.
27. Tipler, p. 71.
28. Hoyle and Wickramasinghe, *Evolution From Space*.
29. Ibid., p. 148, 24, 150, 30, 31.
30. Ibid., p. 143, 130, 130, 96, 96, 130.
31. Ibid., p. 33.
32. Ibid., p. 130, 147.
33. Ibid., p. 51.
34. Ibid., p. 31, 139.
35. Ibid., p. 126.
36. Ibid., p. 127.
37. Ibid., p. 127.
38. Ibid., p. 127.
39. J.D. Bernal, 1967. *The Origin of Life*. London: George Weidenfield and Nicholson, p. 141.

40. J.W.N. Sullivan, 1933. *The Limitations of Science*. New York: The Viking Press, 1949. New York: Mentor books, 11th printing 1963. Time quote on back cover of Mentor edition, 1963.
41. Ibid., p. 94.
42. Ibid., p. 94.
43. A. Einstein, quoted in R. Jastrow, 1978. *God and the Astronomers*. New York: W.W. Norton & Company, p. 113.
44. Hans Gaffron, 1960. In Sol. Tax, ed., *Evolution After Darwin*, Chicago: ed. by Sol Tax, University of Chicago Press, vol. 1, p. 45.
45. Louis de Broglie, 1953. *The Revolution in Physics*. New York: Noonday Press, New York, p. 14.
46. Robert Jastrow, *God and the Astronomers*.
47. National Science Teachers Association position statement, April, 1981. "Inclusion of Nonscience Theories in Science Instruction," *The Science Teacher*, p. 33.
48. Simon Laplace, cited in E.T. Bell, 1937. *Men of Mathmatics*. New York: Simon and Schuster, p. 181.
49. L.E. Orgel, 1973. *The Origins of Life*. John Wiley and Sons, Inc., New York, p. 192.
50. Preston Cloud, 1978. In *A Compendium of Information of the Theory of Evolution and the Evolution-Creationism Controversy*, ed. Jerry P. Lightner, printed by the National Association of Biology Teachers, 11250 Roger Bacon Drive, Reston, VA 22090, p. 83.
51. William Stansfield, 1977. *The Science of Evolution*. New York: MacMillan, p. 10.
52. A.N. Whitehead, 1967 (originally published 1925). *Science and the Modern World*. New York: The Free Press, Chapter 1.
53. Melvin Calvin, 1969. *Chemical Evolution*. New York: Oxford University Press, p. 258.
54. M.B. Foster, 1934. *Mind* **43**, 446.
55. R. Hooykaas, 1972. *Religion and the Rise of Modern Science*. Grand Rapids, Michigan: Wm. B. Erdmans.
56. Loren Eisley, 1961. *Darwin's Century: Evolution and the Men Who Discovered It*. Garden City, New York: Doubleday, Anchor, p. 62.
57. C.F. von Weizsacker, 1964. *The Relevance of Physics*. New York: Harper and Row, p. 163.
58. Stanley Jaki, 1974. *Science and Creation*. Edinburgh and London: Scottish Academic Press.
59. J. Robert Oppenheimer. *Encounter*, October, 1962.
60. Langdon Gilkey, 1959. *Maker of Heaven and Earth*. Garden City, New York: Doubleday, Anchor, p. 9, 125, 129ff.
61. Eisley, p. 62.
62. Hilde Hein, 1971. *On the Nature and Origin of Life*. New York: McGraw-Hill, p. 93.
63. Ibid., p. 93.
64. Michael Polanyi, 1958. *Personal Knowledge*. New York: Harper and Row.
65. Karl Popper, 1959. *The Logic of Scientific Discovery*. New York: Basic Books; 1962. *Conjectures and Refutations: The Growth of Scientific Knowledge*. New York: Basic Books.
66. Thomas S. Kuhn, 1970. *The Structure of Scientific Revolutions*. 2nd ed. Chicago: University of Chicago Press.
67. Stephen Toulmin, 1963. *Foresight and Understanding*. New York: Harper Torchbook.
68. D. Bohm, p. 41.

69. Ibid., p. 42.
70. Ian G. Barbour, 1960. In *Science Ponders Religion*. New York: Appleton-Century-Crofts, p. 200.
71. George Wald, 1979. "The Origin of Life," in *Life: Origin and Evolution*, with Introductions by Clair Edwin Folsome. San Francisco: W.H. Freeman, p. 47.
72. Clair Edwin Folsome, 1979. Introductin to *Life: Origin and Evolution*. San Francisco: W.H. Freeman, p. 2-4.
73. Ibid., p. 3.
74. Ibid., p. 3.
75. P. Fong, 1973. In *Biogenesis, Evolution, Homeostasis*. Ed., A. Locker. New York: Springer-Verlag, p. 93.
76. H.P. Yockey, 1981. *J. Theoret. Biol.* **91**, 13.
77. A.E. Wilder Smith, 1970. *The Creation of Life*. Wheaton, Illinois: Harold Shaw Publishers, 161ff.
78. Melvin Calvin, p. 252.
79. P. William Davis and E. Pearl Solomon, 1974. *The World of Biology*. New York: McGraw-Hill, p. 414.
80. J. Bergman, 1979. *Teaching about the Creation/Evolution Controversy*, Fastback No. 134, Phi Delta Kappa Educational Foundation, Bloomington, Indiana.
81. Charles Darwin, 1963. *The Origin of Species*, with Introduction by Hampton L. Carson. New York: Washington Square Press, Darwin's Introduction, XXXLV.

APPENDIX 1

An Alternative Calculation of the Total Work of Protein Formation

In Chapter 8 the number of unique or distinguishable polymer sequences, Ω_c, was calculated using eq. 8-7. An alternative but equivalent approach presented by Brillouin[1] and Yockey[2] is to consider the number of different symbols that might be incorporated into each position, with the total number of sequences being the product of the number of symbols times the number of positions in the sequence. The result then is

$$\Omega_c = i^N \qquad \text{(App. 1-1)}$$

where typical values for i and N have previously been given (Chapter 8).

This relationship requires the assumption that each of the i symbols is equally probable. A similar relationship can be derived which allows for symbols of different rather than equal probability.[3] The number of sequences predicted by eq. App. 1-1 will always be larger than that predicted by eq. 8-7, since it allows for many different sets of $n_1 + n_2 + n_3 ... + n_i = N$ rather than a given set of n_i values which one could substitute in eq. 8-7. In fact, it can be shown that if one were to evaluate eq. 8-7 for each possible set of $n_1 + n_2 + n_3 ... + n_i = N$ values and sum these results, the total would be identical to that given directly by eq. App. 1-1.

Consider a hypothetical protein of 100 amino acids of 20 types (N = 100, i = 20) and assume that an equal number of each of the 20, i.e., 5, are present in this protein. Using eq. 8-7 we may calculate the number of distinctive sequences for this set of amino acids to be 1.28×10^{115}. If we allow the number of each type of amino acid to assume any value in the range of 0-100, as long as the sum $20n_i = 100$ (i = 1) is retained, additional distinctive sequences are possible. The 1.28×10^{115} sequences possible for $n_1 = n_2 = \ldots n_{20} = 5$ would be added to additional distinctive sequences—for example, for $n_1 = 3$, $n_2 = 7$, $n_3 = n_4 \ldots n_{20} = 5$, and all other possible combinations of n_i. The sum of all these distinctive sequences is calculated using eq. App. 1-1 which gives

$$\Omega_c = i^N = 20^{100} = 1.26 \times 10^{130}$$

Yockey[4] has done a more rigorous analysis for cytochrome c, a protein found in different animals (with somewhat different structures for each cytochrome c, we might add). He modifies eq. 8-7 to allow for an unequal probability of occurrence of each amino acid, based on observed frequencies of appearance in actual proteins. He calculates the number of distinctive sequences of 101 amino acids to be 1.8×10^{126}, a number which is bracketed by our two previous estimates of 1.28×10^{115} using eq. 8-7 and 1.26×10^{130} using eq. App. 1-1. We may be sure that eq. 8-7 gives a lower bound to the number of distinctive sequences observed in a given polypeptide, given that it restricts consideration to the set on n_i values observed in the specified-sequence polypeptide, or protein. Therefore, eq. 8-7 will be used throughout the remainder of this book as a lower bound estimate of Ω_c.

References for Appendix 1

1. L. Brillouin, 1951. *J. Appl. Phys.* **22**, 338.
2. Hubert P. Yockey, 1977. *J. Theoret. Biol.* **67**, 377.
3. C. Shannon, 1948. *Mathematical Theory of Communications.* Urbana: The University of Illinois Press.
4. Yockey, *J. Theoret. Biol.* **67**, 345.

Selected Readings

Bernal, J.D., *The Origin of Life,* Weidenfeld and Nicholson, London, 1967.

Blum, H.F., *Time's Arrow and Evolution,* Harper and Row Publishers Inc., New York, 1962.

Brooks, J. and Shaw, G., *Origin and Development of Living Systems,* Academic Press, New York, 1973.

Day, William, *Genesis on Planet Earth,* House of Talos Publishers, East Lansing, Michigan, 1979.

Folsome, C.E., *The Origin of Life,* W.H. Freeman and Company, San Francisco, 1979.

Fox, S.W. and Dose, K., *Molecular Evolution and the Origins of Life,* Marcel Dekker, 1977.

Kenyon, D.H. and Steinmen, G., *Biochemical Predestination,* McGraw-Hill Co., New York, 1969.

Miller, S.L. and Orgel, L. *The Origins of Life on Earth,* Prentice-Hall, Inc., Englewood Cliffs, New Jersey, 1974.

Ponnamperuma, C., *The Origins of Life,* Dutton, New York, 1972.

Rutten, M.G., *The Origin of Life,* Elsevier Publishing Co., Amsterdam, 1971.

Smith, A.E. Wilder, *The Creation of Life,* Harold Shaw Publishers, Wheaton, Illinois, 1970.

Index

ABOUT THE AUTHORS

Charles B. Thaxton received his Ph.D. in chemistry from Iowa State University. He was a postdoctoral Fellow at Harvard for two years where he studied the history and philosophy of science. He had a postdoctoral appointment for three years in the molecular biology laboratory at Brandeis University. Dr. Thaxton holds memberships in the American Chemical Society, the AAAS, and is a Fellow of the American Scientific Affiliation. He is currently Director of Curriculum Research of the Foundation for Thought and Ethics, Dallas, Tx.

Walter L. Bradley received his Ph.D. in materials science from the University of Texas. He has participated as principal or co-principal investigator on over a million dollars of contract research, and has consulted for many major corporations. He has published over 30 research papers in refereed journals. He has held the Texas Engineering Experimental Station Research Fellowship since 1982 and is Professor of Mechanical Engineering at Texas A&M University.

Roger L. Olsen received his Ph.D. in geochemistry from Colorado School of Mines. He served for two years as Senior Research Chemist with Rockwell Int'l. He has published several technical papers, and has written over 40 confidential engineering/scientific reports. Dr. Olsen holds memberships in Sigma Xi and the American Chemical Society. He is currently Project Supervisor for D'Appolonia Waste Management Services, Englewood, CO.